The Fundamentals of Algorithmic Processes

THE FUNDAMENTALS OF ALGORITHMIC PROCESSES

Sourabh Pal

ARCLER
PRESS

www.arclerpress.com

The Fundamentals of Algorithmic Processes

Sourabh Pal

Arcler Press
224 Shoreacres Road
Burlington, ON L7L 2H2
Canada
www.arclerpress.com
Email: orders@arclereducation.com

ISBN: 978-1-77469-435-0 (Hardcover)

ABOUT THE AUTHOR

Saurabh Pal received his M.Sc. in Computer Science in 1996 and obtained his Ph.D in 2002. He then joined the Department of Computer Applications, VBS Purvanchal University, Jaunpur as a Lecturer. Currently, he is working as Professor. He has authored more than 100 research papers in SCI/Scopus in international/national conference/journals as well as authored four books and also guided many research scholars in computer science/applications. He is an active member of CSI, Society of Statistics and Computer Applications and working as editor, member of editorial board for more than 15 international journals. His research interests include bioinformatics, machine learning, data mining, and artificial intelligence.

TABLE OF CONTENTS

LIST OF FIGURES

LIST OF TABLES

LIST OF ABBREVIATIONS

ACO	ant colony optimization
AMP	adaptive mesh problem
AS	approximation scheme
B2C	business-to-consumer
CNF	conjunctive normal form
CSR	corporate social responsibility
EAs	evolutionary algorithms
ERM	empirical risk minimization
EU	European Union
FMO	Fenna-Matthews-Olson
FQ-Tree	fixed query tree
Gas	genetic algorithms
GRAPE	gradient ascent pulse engineering
GUIs	graphical user interfaces
HL	Hamiltonian
ISP	internet service provider
NP	nondeterministic polynomial time
PETs	privacy-enhancing technologies
PSO	particle swarm optimization
PTAS	polynomial-time approximation scheme
RAM	random-access machine
SOFM	self-organizing feature map
SOMs	self-organized maps
SRM	structural risk minimization
SVM	support vector machine
VPNs	virtual private networks
VP-Tree	vantage point tree

PREFACE

This book provides a comprehensive summary of contemporary research on computer algorithms. The book contains a detailed discussion of several sorts of computer algorithms. The information on algorithm creation and analysis is meant for a broad audience. A fundamental understanding of programming languages and mathematics are required to conduct algorithmic analysis. Knowledge of algorithms enables us to focus on the difficult task of solving a particular problem, rather than on the technical aspects of instructing a computer to perform a particular task. The purpose of this book on algorithms and data structures is to acquaint readers with the theoretical underpinnings of the abilities required to develop computer programs and algorithms. This book is an attempt to familiarize readers with a variety of related fields, such as algorithmic complexity and computability, which should be studied in conjunction with developing applied programming skills.

Typically, the book is comprised of self-contained material that demonstrates a comprehensive understanding of core programming and mathematical concepts. The book's fundamentals are predicated on the introduction of algorithms and data structures in relation to various algorithmic issues. The book explores the many sorts of algorithms available for issue solving. I believe in associate learning, which entails associating one subject with another, i.e., one subject leads to another, and so on. This book comprises topics that are inextricably related. The intention was not to create a comprehensive compendium of everything known about algorithms, but rather to provide a collection of fundamental ingredients and key building blocks upon which algorithms can be built.

Grasp algorithmic challenges require a solid understanding of the underlying concepts of algorithms and data structures. Chapter 1 of the book provides a thorough overview of the fundamental concepts of algorithms and data structures. Algorithms of many types are being researched at the moment. Chapter 2 discusses the categorization of many sorts of algorithms. At the moment, substantial research is being conducted on the subject of search algorithms. Chapters 3 and 4 offer in-depth discussions of complex algorithms such as algorithmic search and quantum walk. On the other hand, Chapter 5 discusses the essential ideas and forms of heuristic algorithms.

In the real-time environment, several algorithms are used to tackle various types of issues. For instance, machine learning algorithms are used to investigate the problems of supervised and unsupervised learning, as discussed in Chapter 6. Chapter 7 discusses the fundamentals of approximation algorithms. Every field of science and technology is governed by some rules. Similarly, algorithmic systems are also governed by certain rules and regulations. Chapter 8 discusses the analytical framework of algorithmic governance and the potential risks and limitations associated with the algorithms and their applicability.

The reader of this book is meant to get knowledge about known ways of successfully resolving difficulties. They will become acquainted with various cutting-edge data structures and novel methods for utilizing data structures to improve the efficacy of algorithms. Because the book is virtually self-contained, it may be used as a course book, reference book, or self-study resource.

—Author

CHAPTER 1

FUNDAMENTALS OF ALGORITHMS

CONTENTS

1.1. INTRODUCTION

We describe an algorithm like any well-described computational process that accepts every value, or set of values, as input and outputs any value or set of values. As a result, an algorithm is a set of computing procedures that changes any input into an output. A further way to think about an algorithm is as a tool for solving a well-defined computing issue (Aggarwal and Vitter, 1988; Agrawal et al., 2004). Generally, the issue description presupposes the intended relationship between outputs and inputs, but the algorithm specifies a specific computing procedure for attaining that relationship. Consider the following scenario: we wish to sort a series of numbers into ascending order. This issue occurred often in practice, providing fertile ground for the presentation of a wide range of conventional design approaches and analytical tools (Hall et al., 1962; Abramowitz and Stegun, 1965). The following is how we formalize any ordering problem:

- **Input:** A sequence of numbers (a_1, a_2, …, a_n).
- **Output:** A permutation (rearrangement) $< \pm_1^{\ddot{u}}, {}_2, \ldots, {}_n >$ of the input arrangement in a way that $\pm_1^{\ddot{u}} \leq {}_2 \leq \cdots \leq {}_n$.

Consider the following input arrangement as an instance (30, 35, 59, 26, 35, 57). The (26, 30, 35, 35, 57, 59) sequence would be returned as the result of a sorting algorithm. An example of the sorting problem is this kind of input arrangement. Generally, a problem example has an input (that satisfies any restrictions imposed by the statement of the problem) that is necessary to compute a solution to the encountered problem.

Sorting is considered a basic step in computer science since it is helpful as an intermediary step in many applications. As a result, we now have access to a huge variety of effective sorting algorithms (Ahuja and Orlin, 1989; Ahuja et al., 1989). The best algorithm for any specific application is determined by various factors, including the different items to be organized, the degree whereby the items are slightly organized, the computer architecture, potential item value limitations, and the kind of storage devices to be utilized: disks, main memory, or tapes.

When an algorithm halts with the appropriate output for each input example, it is thought to be appropriate. A proper algorithm solves the stated computing issue. An erroneous algorithm, on either side, cannot halt at all on certain input examples, or it can halt with an inaccurate response (Ahuja et al., 1989; Courcoubetis et al., 1992). In contrast to our assumptions, erroneous algorithms may occasionally be advantageous if the rate of mistake may be controlled. Therefore, we shall only be concerned with accurate algorithms (Figure 1.1) (Szymanski and Van Wyk, 1983; King, 1995; Didier, 2009).

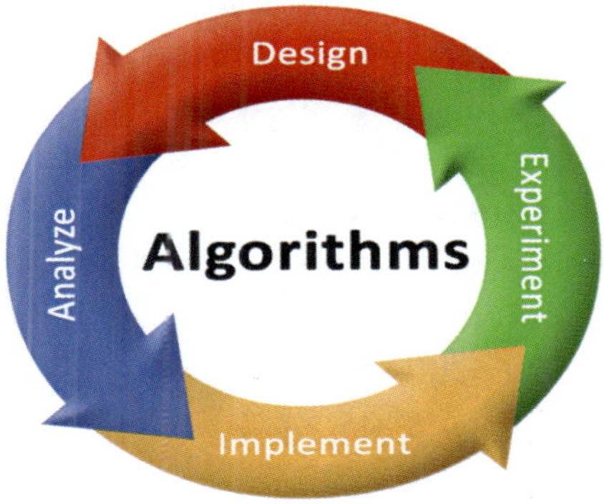

Figure 1.1. Algorithms' basic structure and overview.

Source: http://openclassroom.stanford.edu/MainFolder/CoursePage.php?course=IntroToAlgorithms.

An algorithm may be written in English, in the format of a computer program, or even as a hardware design. The only criterion is that the problem definition must include a clear description of the computational procedure to be used (Snyder, 1984). The following are the features of an algorithm:

- Each instruction should take an indefinite amount of time to complete;
- Almost every instruction should be explicit and accurate, that is, each instruction must have just one meaning;
- The consumer must get the expected outcomes after following the directions;
- There must be no unlimited repeating of a single or several instructions, implying that the algorithm must eventually finish.

1.2. VARIOUS TYPES OF ISSUES SOLVED BY ALGORITHMS

Scientists have devised algorithms for a variety of computer problems, including sorting. Algorithms have a broad range of practical usage across the world. The following are certain such instances:

- The Human Genome Project are made tremendous progress in recent years toward the goal of classifying all 100,000 genes found in the DNA of humans, specifying the sequences of the 3 billion chemical base pairs that make up human DNA, storing this massive amount of information in databases, and developing data analysis techniques. Each procedure necessitates the use of complex algorithms. Several strategies for solving such biological challenges use various principles, allowing scientists to complete jobs while maximizing resource use (Aki, 1989; Akra and Bazzi, 1998). We may obtain more data from laboratory procedures, which saves us time, both machine and human, and money (Regli, 1992; Ajtai et al., 2001).

- The Internet allows individuals worldwide to swiftly access and recover vast amounts of data. Various websites on the Internet handle and utilize this massive amount of data with the help of intelligent algorithms. Identifying effective paths for data to go and utilizing a search engine that may rapidly identify sites with our exact necessary information are two instances of challenges that fundamentally make utilization of various algorithms (Alon, 1990).
- It allows us to negotiate and electronically trade services and things, which is a major advantage of electronic commerce. It is reliant upon the confidentiality of personal data such as details of credit card, keywords, and statements of bank to function properly. Public-key cryptography and digital signs are two of the essential technologies which are used in electronic commerce, and they are both important. Numerical algorithms and numeral theory are the foundations of such technologies (Andersson and Thorup, 2000; Bakker et al., 2012).
- Manufacturing and a variety of other commercial activities occasionally demand the allocation of scarce sources in the most advantageous manner. For example, an oil firm would want to determine where it must locate its wells to maximize the amount of profit it expects to make. Assigning crews to flights in the most cost-effective way feasible while ensuring that every trip is covered and that all applicable government laws regarding the scheduling of the crew are met is something an airline would aspire to do (Ramadge and Wonham, 1989; Amir et al., 2006). Certain political candidates may be faced with the decision of where to spend their money when purchasing campaign advertising to

increase their chances of winning the election. An ISP (internet service provider) may be interested in knowing where they might put extra resources to better satisfy their customers' needs. All of the issues are instances of problems that may be handled by applying linear programming techniques (Figure 1.2) (Dengiz et al., 1997; Berger and Barkaoui, 2004).

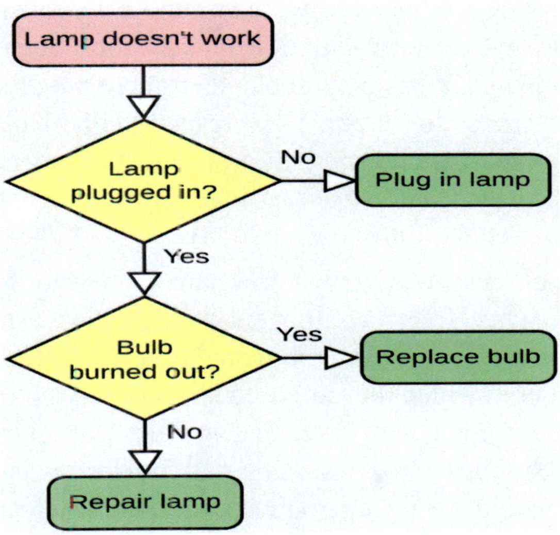

Figure 1.2. A distinctive illustration of algorithm application.

Source: https://invisiblecomputer.wonderhowto.com/how-to/coding-fundamentals-introduction-data-structures-and-algorithms-0135613/.

We would only discuss the basic approaches that imply to such difficulties and problem regions. We will look at ways to tackle a variety of specific issues, such as the ones listed below:

- Assume we have a detailed plan with the distance among each set of two intersections noted, and we want to discover the quickest path from one intersection to the next. Even if we ban routes from crossing over themselves, there might be a tremendous number of viable paths (Festa and Resende, 2002; Zhu and Wilhelm, 2006). How do you pick the fastest of all the available paths in this scenario? In this case, we will describe the plan as a graph and then seek the shortest path between the graph's vertex points.
- Suppose that we have two systematic orders of symbols, $X = (x_1, x_2, \ldots, x_m)$ and $Y = (y_1, y_2, \ldots, y_n)$ as well as we want to discover

the largest communal subsequence of Y and X. A subsequence of X would comprise X minus with some (or all) of its components removed. One subsequence of (A, B, C, D, E, F, G, H, I) is, for example (B, C, E, G). The length of shared subsequence of X and Y determines the degree of similarity between the two sequences. For instance, if the two orders under consideration are basic pairs in the strands of DNA, we can choose to consider them same because we share a lengthy ordinary subsequence. If X has m symbols and Y has n symbols, then there are 2^m and 2^n possible subsequences for X and Y, correspondingly. Unless both m and n are very tiny, choosing all probable subsequences of X and Y and then by matching it can take an excessive amount of time (Wallace et al., 2004; Wang, 2008).

- We are presented with a mechanical design for a library of components. Every component may include examples of other components, and we are obligated to mention the components in the order in which they are used so that every component appears before any other component that includes that component (Maurer, 1985; Smith, 1986). Assuming that the design has n pieces, then there would be nŠ alternative ordering, where the number nŠ represents the factorial function of the design. Since the factorial function increases more quickly as compared to an exponential function, this is not viable for us to produce every potential order and afterwards carry out its validation inside that order in such a way that every component comes before the components that are utilizing it. A problem of this nature is an example of topological sorting (Herr, 1980; Stock and Watson, 2001).
- In this case, let us suppose we have been given n points in the plane, and we want to discover the convex hull of such points, which is the small convex polygon encompassing the points in the plane. Immediately, we may believe of every point as being distinguished by the presence of a nail penetrating through a piece of cardboard. The convex hull will be indicated through a tight rubber band that surrounds all of the nails at this point. In the convex hull, every nail about which the rubber band makes a turn is represented by one of its vertices. As a result, every of the 2n subsets of the points may be used as the vertices of the convex hull. Knowing the points that are vertices of the convex hull is not sufficient in this case; we also require to be familiar with the

sequence in which they occur on the convex hull. Because of this, a variety of options for the convex hull vertices are accessible to the user (Price, 1973; Berry and Howls, 2012).

The above lists have the following characteristics with a wide range of exciting algorithmic problems:

- They have many potential solutions, a large percentage of which fail to fix the problem. Identifying the "best" solution or the one that addresses the problem might be difficult (Arora, 1994).
- The majority of them have real-world applicability. Finding the shortest route was the simplest of the issues described above. Any transport organization, including a shipping company or a railroad, has a financial incentive to locate the shortest route via a road or a rail network because shorter routes result in cheaper labor and fuel costs (Bellare and Sudan, 1994; Friedl and Sudan, 1995). A routing node on the Internet, for example, could want to find the quickest way by the network so that a message can be routed quickly. Alternatively, someone going from Washington to Chicago may choose to get driving directions from a reliable website or use his or her GPS while driving (Sudan, 1992; Arora and Lund, 1996).
- It isn't required for every issue to be addressed by algorithms because a collection of contender solutions may be easily discovered. Consider the case when we are given a set of numerical figures that show signal specimens. We'd want to calculate the discrete DFT of such data right now (Fredman and Willard, 1993; Raman, 1996). The DFT translates the domain of time into the domain of frequency by creating a set of numeric coefficients, allowing us to specify the intensity of distinct frequencies in the sampled signal. Apart from being at the heart of signal processing, DFT has extensive applications in the compression of data, and multiplication of huge integers and polynomials (Polishchuk and Spielman, 1994; Thorup, 1997).

1.3. DATA STRUCTURES

A data structure is a method of storing and arranging information to make it easier to modify and access. No one data structure generally works fine for all applications, therefore it's significant to understand the benefits and drawbacks of some of them (Brodnik et al., 1997).

1.3.1. Hard Problems

Talking When it comes to efficient algorithms, the most commonly used metric for efficiency is speed, which is defined as the amount of time it takes for an algorithm to output its consequence. Furthermore, there have been a few situations for which there is ineffective solution at any given point in time. It is considered worthwhile to study problems with NP (nondeterministic polynomial time) solutions for two main reasons (Phillips and Westbrook, 1993; Ciurea and Ciupala, 2001). Firstly, even if there has not yet been discovered an efficient algorithm for an NP-complete issue, no single individual has ever demonstrated that an efficient algorithm for this problem can't be discovered. Nobody knows whether or whether not there are any effective algorithms available for NP-complete problems, which is a true statement. As a second point, there is an interesting feature of a collection of NP-complete problems in that if an efficient algorithm emerges for any one of them, then such an effective algorithm would exist for all of them as well (Cheriyan and Hagerup, 1989, 1995). Because of the link among NP-complete issues and the absence of effective solutions, the lack of effective solutions becomes even more frustrating. Third, multiple NP-complete problems are comparable to each other, but they are not similar to the issues for which we already have effective algorithms. Computer professionals are fascinated by the fact that a slight modification in the issue description may result in a significant difference in the competency of the most well-known algorithm in the world (Cheriyan et al., 1990, 1996).

It is important to be conversant with NP-complete issues since certain of them emerge unexpectedly frequently in real-world applications. Given the fact that you have been tasked with developing an effective algorithm for any NP-complete issue, it is likely that you would waste a significant amount of time in an unproductive pursuit. If on either side, you may demonstrate that the issue is NP-complete, you may devote your attention instead to inventing a proficient algorithm that delivers a decent, but the inefficient solution (Leighton, 1996; Roura, 2001; Drmota and Szpankowski, 2013).

Consider the case of a delivery company with a main depot, which is already in existence. Throughout the day, its delivery vehicles are loaded at the depot and then dispatched to different locations to deliver items to customers. Every truck is expected to return to the depot before the end of the day so that it may be prepared for loading the following day's loads (Bentley et al., 1980; Chan, 2000; Yap, 2011). Specifically, the corporation desires to pick a series of delivery stops that results in the shortest total distance traveled by every vehicle to reduce prices. In this issue, which is known as the "traveling-salesman problem," the NP-completeness of the solution has been established. It doesn't have a fast algorithm to work with. Furthermore, given some specific assumptions, we may identify effective algorithms that provide an average distance that is not more than the shortest feasible distance (Shen and Marston, 1995; Verma, 1997).

1.3.2. Parallelism

Over many years, we had been able to depend upon the speed of processor clock increasing at a consistent pace. Even though physical restrictions provide an eventual impediment to ever-increasing clock speeds, the following is true: as with the clock's speed, power density increases super-linearly; as a result, chips are at risk of melting once their clock speeds reach a certain threshold (Meijer and Akl, 1987, 1988). We build circuits with many processing cores to conduct more computations per second and so achieve higher throughput. Such multicore computers may be compared to a large number of sequential computers on a single chip, or we may refer to them as a form of "parallel computer," depending upon your perspective. To get the maximum potential performance out of multicore machines, we should design algorithms with parallelism in mind while developing them. Algorithms that use several cores, like "multithreaded" algorithms, are particularly advantageous. From a theoretical standpoint, this sort of model has significant advantages, and as a result, it serves as the foundation for a large number of successful computer systems (Wilf, 1984).

1.4. ALGORITHMS LIKE A TECHNOLOGY

Assume that computers had been significantly faster than they were now and their memory had been completely free. May you imagine every other cause to be interested in algorithms at this point? We will still like to explain that our solution ends and does so with an accurate result, therefore the answer remains affirmative. Given the high speed with which computers operate, any exact approach of problem-solving would be rendered ineffective (Igarashi et al., 1987; Park and Oldfield, 1993). You most likely will want your software execution to fall inside the limits of good software engineering practice (for instance, your execution must be well recorded and well designed), although, you are more than likely going to use whatever method is the quickest and most straightforward to put into action (Glasser and Austin Barron, 1983).

Computers can be extremely fast, but they are not capable of becoming indefinitely fast. Similar to that, while their memory can be cheap, it's not completely free. Time for computation is, thus, a finite source, just as memory is a finite resource. It is important to make effective use of such resources, and algorithms that are effective in terms of space or time may assist us in doing so (Wilson, 1991; Buchbinder et al., 2010).

Efficiencies of different algorithms that are created for the solution of the same issue frequently differ significantly in terms of their effectiveness. When compared to the variations resulting from hardware and software, such changes might be far more noticeable (Vanderbei, 1980; Babaioff et al., 2007; Bateni et al., 2010).

For greater understanding, consider two alternate sorting methods. The 1st is known as insertion sort, and it takes about equivalent to c_1n^2 to sort n items; where c_1 is a constant that is independent of n; this implies that it takes roughly proportionate to n^2. The 2nd technique is known as merge sort, and it takes roughly the same amount of time as $c_2$nlgn, wherein lgn represents $\log_2$n and c_2 is another constant that is independent of n. Insertion sort often is a small constant variable as compared to merge sort, like c_1 is smaller than c_2. We'll see that constant variables have a far less influence on the running duration than the reliance upon the input size n (Aslam, 2001; Du and Atallah, 2001; Eppstein et al., 2005). Let's write insertion sort's running time as c_1n.n and merge sort's running time as c_2n.lg n. Then we see that where insertion sort has a factor of n in its running time, merge sort has a factor of lg n, which is much smaller.. Even though insertion sort is normally quicker than merge sort for smaller input sizes, once the size of input n grows large enough, merge sort's benefit of lgn versus n would greater than compensate for the dissimilarity in constant variables. There would always be a crossover point outside of which merge sort is quicker, irrespective of how much smaller c_1 is relative to c_2 (Sardelis and Valahas, 1999; Dickerson et al., 2003).

Consider the following scenario: a faster computer (designated as computer A) is conducting insertion sort besides a slower computer (designated as computer B) that is doing merge sort. Every one of them should sort a 10 million-number array that has been given to them. (Although 10 million digits can sound high, given that the numbers are 8-byte integers, the input would only take up around 80 megabytes, which is more than enough space for even a low-cost laptop computer's memory to be filled several times over.) Take, for example, the assumption that computer A is capable of processing 10 billion instructions for one second, while computer B is capable of processing just 10 million instructions per second; this makes computer A 1,000 times more powerful than computer B in terms of raw computing capacity (Goodrich et al., 2005). Let us assume that the world's most cunning programer writes insertion sort for computer A in machine language, and the resultant code needs $2n^2$ instructions to arrange n integers, to emphasize the drastic difference. Furthermore, consider the following scenario: an average programer applies merge sort by utilizing a higher-level language and an ineffective compiler, and the resultant code requires 50nlgn instructions. Thus, computer A would require the following resources to sort 10 million numbers:

$$\frac{2\cdot(10^7)\,instructions}{10^{10}\ instructions/\sec ond} = 20,000 \sec ond (more\ than\ 5.5\ hours)$$

On the other hand, computer B would take:

$$\frac{50\cdot10^7 \lg 10^7\ instructions}{10^7\ instructions/\sec ond} \approx 1163 \sec ond (more\ than\ 20 \min utes)$$

Yet with a poor compiler, computer B would be greater than 17th times quicker than computer Through utilizing an algorithm that running time increases more gradually. When sorting 100 million values, the benefit of merge sort becomes even more apparent: insertion sort would take more than 23 days, but merge sort would take less than 4 hours. Generally, the relative advantage of merge sorting decreases as the size of the problem grows (Eppstein et al., 2008; Acar, 2009).

1.5. ALGORITHMS AND OTHER TECHNOLOGIES

The preceding case demonstrates that algorithms, such as the hardware of computer, must be considered technology. The system's overall performance is determined by both the use of effective algorithms and the use of fast hardware. Algorithms are making fast development in the same way that other computer technologies are (Amtoft et al., 2002; Ausiello et al., 2012). You could wonder if algorithms are genuinely that is significant on modern computers when other sophisticated technologies are taken into account, such as:

- Intuitive, simple graphical user interfaces (GUIs);
- Advanced computer architectures and production technologies;
- Object-oriented systems and Object-Oriented Programming;

- Wireless and wired networking that is quick;
- Web technologies which are integrated.

Yes, it is correct. Although some programs (such as simple web-based apps) may not need algorithmic content at the utilization level, many others do. Suppose a Web-based service that explains how to go from one area to another (Szymanski, 1975; Aslam, 2001). Such service's implementation will be reliant on a graphical user interface, wide-area networking, fast hardware, and, most likely, object orientation. However, algorithms will be required for specific activities like identifying routes (most likely utilizing a shortest-path method), interpolating addresses, and displaying maps (Bach, 1990; Avidan and Shamir, 2007).

Furthermore, even an application that doesn't need algorithmic data at the application level is heavily reliant on algorithms. Because the application requires fast hardware and the design hardware utilize algorithms. Moreover, the program makes utilize GUIs, which are created using algorithms (Winograd, 1970; Bach and Shallit, 1996). Furthermore, the application might depend on networking, and routing in networks is mostly based on algorithms. If the program is written in a language apart from machine language, it should be processed through a compiler, assembler, or interpreter, each of which employs a variety of algorithms. As a result, we may determine that algorithms are at the heart of nearly all computer technology today (Bailey et al., 1991; Baswana et al., 2002).

Furthermore, with computers' ever-increasing capabilities, we are using them to solve more complicated issues than ever before. As may be observed from the comparison of insertion sort and merge sort, the variations in performance among algorithms become more noticeable in the bigger problem (Bayer, 1972; Arge, 2001; Graefe, 2011). One identifying characteristic that distinguishes professional programers from newbies is having a firm foundation of algorithmic knowledge and technique. Since we may do a few things with the aid of contemporary computer technologies without knowing anything about algorithms, we may accomplish a lot more with a solid understanding of algorithms (Blasgen et al., 1977).

1.6. GETTING STARTED WITH ALGORITHM

We begin by taking a look at the insertion sort method, which is used to solve the sorting issue. We shall describe a "pseudocode," which must be recognizable to you if you have any experience with computer programming. It would be used to explain how to express our algorithms, which we shall demonstrate in this section. Following the specification of the insertion sort algorithm, it sorts adequately, and we assess the method's execution time (Kwong and Wood, 1982; Icking et al., 1987). The resulting analysis yields a notation that illustrates how the amount of time required sorting things increases as the number of items to sort items increases. As a conclusion to our discussion of insertion sort, we would present the divide and conquer strategy for algorithm designing and demonstrate how it may be used to construct a new algorithm known as merge sort. Following that, we would conclude with an evaluation of the time required by merge sort to complete its task (Nievergelt, 1974; Beauchemin et al., 1988).

1.6.1. Insertion Sort

The 1st method, insertion sort, solves the sorting issue:

- **Input:** An order of n integers $(a_1, a_2, \ldots, a_n)$ is given as input.
- **Output:** A permutation (rearrangement) $< a_1^{'}, a_2^{'}, \ldots, a_n^{'} >$ of the input arrangement in such a way that $a_1^{'} \leq a_2^{'} \leq \cdots \leq a_n^{'}$.

The keys are the integers that are going to be classified. Even though we are arranging a series theoretically, the input comes in the format of an arrangement with n entries.

Algorithms would generally be described as programs written in pseudocode, which is alike to C++, C, Pascal, Java, or Python in several ways. If you've worked with these languages before, you must have no trouble understanding our algorithms (Figure 1.3) (Monier, 1980; Kim and Pomerance, 1989; Damgård et al., 1993).

Figure 1.3. By using the insertion sort, we arrange the hand of cards.

Source: https://mitpress.mit.edu/books/introduction-algorithms-third-edition.

The distinction among real code and pseudocode is that we specify a particular algorithm using any expressive way is the shortest and simplest. Because English is sometimes the simplest way, it must not be surprising to see a sentence or a phrase in English placed inside a portion of "real" code (Ben-Or, 1983; Bender et al., 2000). Another distinction among real code and pseudocode is that pseudocode is usually unconcerned through software engineering issues. Data abstraction, modularity, and error handling are frequently overlooked to convey the algorithm's spirit more briefly (Wunderlich, 1983; Morain, 1988).

We'll begin with the insertion sort. Insertion sort is a fast algorithm for sorting a limited number of items. We may compare how insertion sort works to how different people arrange a hand of cards. The cards would be dealt face down on the table, starting with an empty left hand. Afterwards, we pick one card at a time from the table and put it in the proper position in the left hand (Fussenegger and Gabow, 1979; John, 1988). To determine the proper placement of a card, we compare it to each of the previously held cards in the hand, from right to left, as indicated in the diagram below. The cards seized in the left hand are categorized in every case. Additionally, these were the cards that were initially at the top of the table's pile (Kirkpatrick, 1981; Munro and Poblete, 1982; Bent and John, 1985).

Our insertion sort's pseudocode is offered as the INSERTION-SORT algorithm, which accepts a sequence as a parameter (A [1, …, n). This arrangement has an n-length sequence that must be sorted. The algorithm then arranges the input array in situ, reordering the integers in array A and storing a constant number exterior to the arrangement as a maximum at any given moment. After the INSERTION-SORT method is completed, the input array A contains the sorted output sequence (Figure 1.4) (Schönhage et al., 1976; Bentley et al., 1980; Bienstock, 2008).

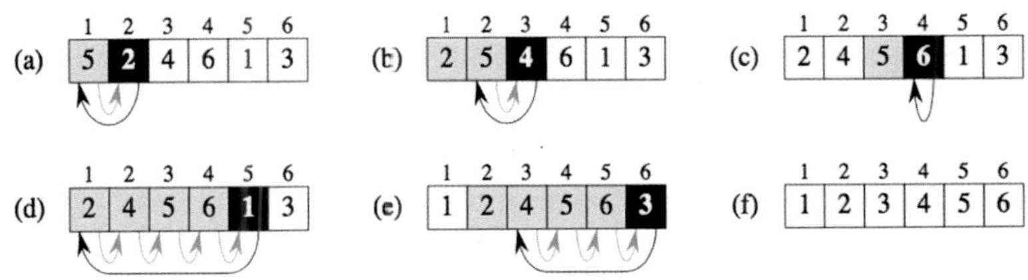

Figure 1.4. On array A, the INSERTION-SORT operation equals (2, 5, 4, 6, 1, 3). The arrangement indices are displayed above the rectangles, while the values stored in the array places are displayed within the rectangles. The reiterations [(a) to (d)] of the 'for loop' of lines 1–8 is shown. The rectangle (black) contains the key retrieved from A(j) throughout every iteration, which is subsequently equated with the values within shaded rectangles to its left inline 5's test. In line 6, shaded arrows show array values that have been moved one place to the right, and black arrows show where the key has moved to in line 8. (e) The most comprehensive sorted array.

Source: https://mitpress.mit.edu/books/introduction-algorithms-third-edition.

The INSERTION SORT (A) is demonstrated below:

1. Key = A[j];
2. For j = 2 to length A;
3. i = j – 1;
4. // Insertion of A[j] into a sorted arrangement a [1, …, j – 1];
5. Whereas A[i] is greater than key and I is greater than zero;
6. A [i + 1] = A[i];
7. // Feeding addition of value in A[i];
8. i = i – 1;
9. A[i + 1] = key.

1.6.2. Loop Invariants and the Precision of Insertion Sort

The mechanism of this technique for A = (2, 5, 4, 6, 1, and 3) is given in Figure 1.4. The "current card" being put into the hand is denoted by index j. The sub-array comprising items A [1, ..., j – 1] establishes the currently sorted hand at the start of every iteration of the "for loop "(indices by j), and the remaining sub-array A[j + 1, ..., n] relates to the heap of cards which are still existing on the table. Elements A [1, ..., j – 1] have been the original elements that are utilized to be in positions 1 through j 1 but are now arranged. Such qualities of A [1, ..., j – 1] are explicitly described like an *invariant of loop.*

To explain why an algorithm is efficient, loop invariants are employed. A loop invariant must effectively verify three things:

- **Initialization:** It is valid before the loop's first iteration.
- **Maintenance:** If it is valid before the loop iteration, it will continue to be accurate before the next iteration.
- **Termination:** As the loop comes to an end, the invariant supplies us with a useful attribute that aids in establishing the correctness of the method.

If the first two characteristics are true, then the invariant of loop is factual before the iteration of the loop. It is important to highlight the similarities between this and mathematical induction. To demonstrate that a characteristic holds, we must show both an inductive step and a base case (Duda et al., 1973; Bloom and Van Reenen, 2002). Establishing that the invariant holds before the 1st iteration correlates to the base situation in this scenario while determining that the invariant holds between iterations relate to the inductive step in this scenario (Dehling and Philipp, 2002; Bienstock and McClosky, 2012).

Because we have been utilizing the invariant of loop to demonstrate accuracy, the 3rd characteristic is likely the most crucial. Usually, we employ the loop invariant in conjunction with the criteria that resulted in the termination of the loop to achieve our goal. When compared to our typical usage of mathematical induction, through which we apply the inductive step indefinitely, in this situation we halt the "induction" once the loop comes to a close (or terminates). Analyze how well these principles hold up in the scenario of insertion sort (Bollerslev et al., 1994).

- **Initialization:** We begin by showing that the loop invariant holds before the first loop iteration when j = 2. As a result, the sub-array

A[1, …, j – 1] is only made up of the single element A[1], which is also the original element in A[1]. Furthermore, this sub-array is ordered, implying that the loop invariant holds before the 1st loop iteration (Williamson, 2002; D'Aristotile et al., 2003).

- **Maintenance:** Now we'll look at the 2nd characteristics: representing that iteration keeps the loop invariant. The for-loop body is working by moves A[j – 1], A[j – 2], A[j – 3], and similarly via single position to the accurate until it locates the correct position for A[j] (lines 4–7), at which point it puts the A[j] value (line 8). The sub-array A[1, …, j], on the other hand, comprises elements that were initially in A[1, …, j], however in a proper array. The loop invariant is conserved by increasing j for each consecutive iteration of the 'for loop' (Nelson and Foster, 1992; Conforti et al., 2010).
- **Termination:** Consequently, we look at what occurs when the loop comes to an end. The condition that causes the 'for loop' to end is that j > A, length = n. We must have j = n + 1 at that point since each loop iteration raises j by 1. By putting n + 1 for j in the loop invariant language, we get the sub-array A[1, …, n], which has the same items as A[1, …, n], but in a proper arrangement. We decide that the complete array is categorized since the sub-array A [1, …, n] is the ideal array. As a result, the algorithm is accurate (Fourier, 1890, 1973).

1.6.3. Pseudocode Conventions

The following conventions are used in our pseudocode:

- The block's structure is shown by indentation. For instance, the body of a 'for loop' starting on line 1 includes lines 2–8, but the body of a while loop starting on line 5 includes lines 6–7 but not line 8. This indentation style can also be used for if-else statements. Utilizing indentation rather than traditional block structure indications like beginning or final statement minimizes clutter whilst maintaining, or even improving, clearness (Verma, 1994, 1997; Faenza and Sanit, 2015).
- Looping structures like 'for loop' and 'while loop,' as well as conditional constructions like if-else and repeat-until, are identical to those found in C, C++, Python, Java, and Pascal. In opposed to several circumstances encountered in C++, Java, and

Pascal, we have supposed that the loop counter keeps its value after exiting the loop. As a result, the counter value of the loop is the value that 1st topped the 'for loop' bound after a 'for loop.' This functionality may be used in our insertion sort accuracy argument (Xiaodong and Qingxiang, 1996).

- In the situation of an if-else statement, the else clause is indented to the same level as its equivalent if clause. Even though we do not use the phrase then, we may refer to the section of the code that is performed when the test is successful if the assertion is true as a then clause. In the situation of multi-ways tests, we take advantage of the else-if condition to run tests after the 1st one has been completed. The majority of block-structured languages have constructs that are similar to one another, even though the actual syntax can be different. In Python, there have been no repetition until loops, and the 'for loops' that do uniquely exist in the language function from the 'for loops' (Blelloch et al., 1994).
- The slash character "//" denotes that the remaining portion of the line is a remark (comment).
- Several assignments of the kind i= j=e that represent the value of expression e to both factors i and j; As a result, it must be viewed as equal to the assignment j= e that came before it, which was itself proceeded through the assignment i= j.
- Factors (like the key, i and j) have been only accessible inside the scope of the given method. It is not permitted to utilize global variables unless they are explicitly stated (Blelloch and Greiner, 1996; Blelloch and Mags, 1996, 2010).
- Accessing items of an array is done by indicating the array name followed by the index given in brackets, as seen in Figure 1.4. For instance, array A[i] signifies the element that is the ith element of array A. The "…" notation denotes a range of values contained within an array of values. In this case, A [1, …, j] represents the sub-array of A having the j elements A[1], A[2], …, A[j].

1.7. ANALYZING ALGORITHMS

We must forecast the resources required by an algorithm during evaluating it. We are often concerned with sources like computer hardware or communication bandwidth, memory, because most of the time, we want to quantify computational time (Frigo et al., 1998; Blumofe and Leiserson,

1999). Identifying the most effective contestant algorithm for a topic is usually done by analyzing multiple contestant algorithms. This form of assessment may reveal more than one feasible candidate; although, we may typically eliminate weaker algorithms throughout the process (Blelloch and Gibbons, 2004).

We need to have an implementation technology model in place before we may perform an algorithm analysis. This model should take into account the technology's resources and its costs. Regarding the implementation technique, we'll use a generic one-processor or the random-access machine (RAM) model of computing. Furthermore, our algorithms would be implemented as computer programs. Every instruction in the RAM model is performed once a time, with no concurrent operations (Brent, 1974; Buhler et al., 1993; Brassard and Bratley, 1996).

The RAM model's instructions, and their charges, should be precisely defined. This effort would not only be arduous but would also provide limited insight into the algorithm's analysis and design. However, extreme caution must be exercised to avoid abusing the RAM model. For instance, if a RAM has a sort instruction, we may arrange with just one instruction. Moreover, such a RAM is impractical as practical computers lack these instructions (Chen and Davis, 1990; Prechelt, 1993). As a result, we shall concentrate on the design of actual computers. The RAM model incorporates data transfer (store, copy, load,), arithmetic (such as division, multiplication, addition, subtraction, remainder, floor, ceiling, and so on), and control instructions that are commonly found in actual computers (subroutine call and return conditional and unconditional branch). Every instruction requires a set amount of time to complete.

1.7.1. Order of Growth

We may improve our study of the INSERTION-SORT easier by using a few simplifying assumptions. The real cost of every statement may be ignored as a 1st step by utilizing the constants c_i to denote such costs. Afterwards, we'll see that even such constants supply us with more information than we require. The worst-case running time was defined in the preceding sections as $an^2 = b\,n + C$ for certain constants a, b, and c that is based on the statement costs c_i. As a result, we may disregard both the abstract costs c_i and the real statement costs.

The sequence or rate of increase of the running time, which we are interested in, is another simplifying assumption that may be created. As a result, we only analyze the leading component of the formula (which is an^2), because the lower-order terms are fairly irrelevant for high values of n. In addition, the constant coefficient of the leading term is ignored since it is less relevant as compared to the rate of increase in finding computing effectiveness for large inputs. In the situation of insertion sort, we only have the n^2 factor from the leading word after discarding the constant coefficient of the leading term and the lower-order terms.

1.7.2. Insertion Sort Analysis

The time needed by the INSERTION-SORT method is mostly dependent on the input: sorting 1,000 numbers, for example, would take significantly longer than sorting three numbers. Furthermore, based upon how much each input sequence has been sorted, INSERTION-SORT might take a varied amount of time to arrange two input sequences of the similar size. Usually, the time needed by an algorithm grows in proportion to the size of the input, thus it is common to express program execution time as a function of input size. To do so, we need to be more careful in our definitions of "size of input" and "runtime."

The optimum estimate for the amount of the input is based upon the problem under analysis. The most common measure for a variety of issues, like sorting or calculating DFT, is the number of items in the input, for instance, n is the size of the array for arrangement. The whole number of bits necessary to indicate the input in conventional binary notation is the most ideal measure of the size of the input for various additional issues, such as the multiplication of two numbers. Sometimes, it is more appropriate to use two integers rather than one to express the input size. For instance, if an algorithm's input is a graph, the number of vertices and edges in the graph may be used to define the size of the input. For each of the difficulties we've looked at, we'll note to you which input size measures are being used.

The primitive operations or the number of executed steps determines an algorithm's *running time* for a given input. It is necessary to define the step idea in such a way that it is machine-independent. Let us adopt the following viewpoint for the time being. A fixed sum of time is needed to implement every line of our pseudocode. Even though the time necessary to execute one line can vary from the time needed to execute another, we would suppose that every implementation of the ith line takes a constant amount of time equivalent to c_i. Such viewpoint is consistent with the RAM model, and it also illustrates how pseudo is implemented on the vast majority of real computers.

The next demonstration looks at the progression of the equation that is utilized to calculate the INSERTION-SORT running time. From a complicated formula that makes usage of all of the statement costs (that is, c_i) to a simplified representation that is brief and easier to manipulate, the expression progresses through the stages. This straightforward language makes it easier to measure the effectiveness of a specific algorithm in comparison to other algorithms. The demonstration of the INSERTION-SORT method in terms of the "cost" associated with every statement time and the number of times it is attempted to execute the statement.

For every j= 2, 3, 4, …, n, wherein n = *length*. A denotes the number of tries the 'while loop' experiment in line 5 is performed for that specific value of j. A 'for' or 'while' statement that exits conventionally (that is, because of a test in a looping header) has its execution performed once more than the body of the loop that exited conventionally. It is considered that comments are non-executable statements that don't spend any time when they are executed.

Sl. No.	Times	Cost	INSERTION SORT (A)
.1	n – 1	0	//Insertion of A[j] into an arranged sequence a[1, …, j – 1]
.2	n – 1	0	\\feeding addition of value in A[i]
.3	n	c1	for j=2 to length.A
.4	n – 1	c2	key = A[j]
.5	n – 1	c4	i = j – 1
.6	$\sum_{j=2}^{n} t_j$	c5	while A[i]>key and i>0
.7	$\sum_{j=2}^{n} (t_j - 1)$	c6	A[i + 1] = A[i]
.8	$\sum_{j=2}^{n} (t_j - 1)$	c7	i = i – 1
.9	n – 1	c8	A[i + 1] = key

The phenomenon has several complexities. In most cases, computational steps defined in English are versions of a technique that takes longer than the expected (set) amount of time. For example, "the sorting of the points by x-coordinates" usually takes longer than the time allocated. It's also worth noting that a statement that invokes a subroutine spends the same amount of time, even if the function takes longer. As a result, for this sort of phenomenon, we identify the calling process.

The overall running time of an algorithm is the addition of the running times of all the statements that have been performed. Every statement which is executed in c_i steps and is executed n times in total will contribute $c_i n$ to the overall running time. We add the multiplication of the times and cost columns to get the overall running time of INSERTION-SORT on an input consisting n values, indicated by T(n), and we get:

$$T(n) = c_1 n + c_2(n-1) + c_4(n-1) + c_5 \sum_{j=2}^{n} t_j + c_6 \sum_{j=2}^{n} (t_j - 1) + c_7 \sum_{j=2}^{n} (t_j - 1) + c_8(n-1)$$

Although when dealing with known-size inputs, an algorithm's execution time can be influenced through which input of that size is specified. The best-case situation in INSERTION-SORT, for example, is when the arrangement is already ordered. Whenever the initial value of i is j – 1, we receive A[i]≤key in line 5 for the values of j= 2, 3, …, n. Consequently, $t_j = 1$ for j = 2, 3, …, n and the running time of the best-case is:

$$T(n) = c_1 n + c_2(n-1) + c_4(n-1) + c_5(n-1) + c_8(n-1)$$
$$= (c_1 + c_2 + c_4 + c_5 + c_8)n - (c_2 + c_4 + c_5 + c_8).$$

For constants a and b, this running time may be represented as +b. Because a and b are both based upon the statement costs ci, n is a *linear function.*

The worst-case situation happens when the array is sorted in reverse order, that is, in decreasing order. Then, every element A[j] should be compared to each element in the entire sorted sub-array A[1, …, j – 1], and consequently tj=j for j = 2, 3, …, n. Remember that:

$$\sum_{j=2}^{n} j = \frac{n(n+1)}{2} - 1$$

and

$$\sum_{j=2}^{n} (j-1) = \frac{n(n-1)}{2}$$

The running time of INSERTION-SORT in the situation of worst-case is:

$$T(n) = c_1 n + c_2(n-1) + c_4(n-1) + c_5\left(\frac{n(n+1)}{2} - 1\right) + c_6\left(\frac{n(n-1)}{2}\right) + c_7\left(\frac{n(n-1)}{2}\right) + c_8(n-1)$$

$$= \left(\frac{c_5}{2} + \frac{c_6}{2} + \frac{c_7}{2}\right)n^2 + \left(c_1 + c_2 + c_4 + \frac{c_5}{2} - \frac{c_6}{2} - \frac{c_7}{2} + c_8\right)n - (c_2 + c_4 + c_5 + c_8).$$

For constants a, b, and c, the running time of such worst-case may be represented as $an^2 = bn + c$. Because these constants are dependent on the statement costs c_i, it is a quadratic function of n.

In the Insertion sort scenario, an algorithm's running time is usually set for every specific input; while there have been some "randomized" algorithms whose behavior varies even for a fixed input.

1.7.3. Analysis of Worst-Case and Average-Case

To better understand the insertion sort, we have studied both cases: the best scenario, in which the input arrangement has already been categorized, and the worst situation, in which the arrangement has already been categorized in the opposite direction. All of our attention is focused solely on the worst-case, which is the longest possible running time for every input of size n. The following are the three most important causes for this perspective:

- By calculating the running time for the worst-case of an algorithm, we may obtain an upper bound on the running time of every input. Understanding the upper bound assures that the algorithm would not take much longer than it has been estimated to take. As a result, it removes the need for us to make an informed judgment, and we may safely trust that things would not become much worse.
- The worst-case scenario occurs fairly frequently for certain algorithms, as seen in Figure 1.4. Whenever we check any database to discover certain particular data, we would frequently come into the worst-case scenario of the searching algorithm, which occurs when the data we seek is not there in the information.
- The "average case" is almost always just as dreadful like the "worst-case" scenario. Consider the following scenario: we randomly select n integers and then apply insertion sort on them. Is it possible to estimate the amount of time necessary for identifying where to put element A[j] in sub-array A[1, …, j – 1] using a timer? Usually, 50% of the items included in A[1, …, j –

1] are more than A[j], and 50% are less than A[j]. As a result, we would be inspecting around 50% of the sub-array A[1, …, j – 1], and as a result, tj is approximately j = 2. The running time of the average case is found to be a quadratic function of the amount of the input as the running time of the worst case is shown to be in the same way.

When dealing with a given situation, the capability of average-case evaluation is controlled since it is not always evident what constitutes an "average" input. Most of the time, we want to suppose that almost all inputs of a particular size are equally probable. In actuality, such a supposition can be broken; yet, at times, we can utilize a *randomized algorithm*, which produces arbitrary selections to permit for probabilistic evaluation and to provide results in a predicted running time by using a randomized method.

REFERENCES

1. Abramowitz, M., & Stegun, I. A., (1965). *Handbook of Mathematical Functions with Formulas, Graphs, and Mathematical Table* (Vol. 2172, pp. 1–23). New York: Dover.
2. Acar, U. A., (2009). Self-adjusting computation:(an overview). In: *Proceedings of the 2009 ACM SIGPLAN Workshop on Partial Evaluation and Program Manipulation* (pp. 1–6). ACM.
3. Ahuja, R. K., & Orlin, J. B., (1989). A fast and simple algorithm for the maximum flow problem. *Operations Research, 37*(5), 748–759.
4. Ahuja, R. K., Orlin, J. B., & Tarjan, R. E., (1989). Improved time bounds for the maximum flow problem. *SIAM Journal on Computing, 18*(5), 939–954.
5. Ajtai, M., Megiddo, N., & Waarts, O., (2001). Improved algorithms and analysis for secretary problems and generalizations. *SIAM Journal on Discrete Mathematics, 14*(1), 1–27.
6. Aki, S. G., (1989). *The Design and Analysis of Parallel Algorithms, 1*, 10.
7. Akra, M., & Bazzi, L., (1998). On the solution of linear recurrence equations. *Computational Optimization and Applications, 10*(2), 195–210.
8. Alon, N., (1990). Generating pseudo-random permutations and maximum flow algorithms. *Information Processing Letters, 35*(4), 201–204.
9. Amir, A., Aumann, Y., Benson, G., Levy, A., Lipsky, O., Porat, E., & Vishne, U., (2006). Pattern matching with address errors: Rearrangement distances. In: *Proceedings of the Seventeenth Annual ACM-SIAM Symposium on Discrete Algorithm* (Vol. 1, pp. 1221–1229). Society for Industrial and Applied Mathematics.
10. Amtoft, T., Consel, C., Danvy, O., & Malmkjær, K., (2002). The abstraction and instantiation of string-matching programs. In: *The Essence of Computation* (pp. 332–357). Springer, Berlin, Heidelberg.
11. Andersson, A. A., & Thorup, M., (2000). Tight (er) worst-case bounds on dynamic searching and priority queues. In: *Proceedings of the Thirty-Second Annual ACM Symposium on Theory of Computing* (Vol. 1, pp. 335–342). ACM.

12. Arge, L., (2001). External memory data structures. In: *European Symposium on Algorithms* (pp. 1–29). Springer, Berlin, Heidelberg.
13. Arora, S., & Lund, C., (1996). Hardness of approximations. In: *Approximation algorithms for NP-Hard Problems* (pp. 399–446). PWS Publishing Co.
14. Arora, S., (1994). *Probabilistic Checking of Proofs and Hardness of Approximation Problems.* Doctoral dissertation, Princeton University, Department of Computer Science.
15. Arora, S., Lund, C., Motwani, R., Sudan, M., & Szegedy, M., (1998). Proof verification and the hardness of approximation problems. *Journal of the ACM (JACM), 45*(3), 501–555.
16. Aslam, J. A., (2001). *A Simple Bound on the Expected Height of a Randomly Built Binary Search Tree (Vol.1, pp.1-10).*
17. Ausiello, G., Crescenzi, P., Gambosi, G., Kann, V., Marchetti-Spaccamela, A., & Protasi, M., (2012). *Complexity and Approximation: Combinatorial Optimization Problems and Their Approximability Properties*. Springer Science & Business Media.
18. Avidan, S., & Shamir, A., (2007). Seam carving for content-aware image resizing. In: *ACM Transactions on Graphics (TOG)* (Vol. 26, No. 3, p. 10). ACM.
19. Babaioff, M., Immorlica, N., Kempe, D., & Kleinberg, R., (2007). A knapsack secretary problem with applications. In: *Approximation, Randomization, and Combinatorial Optimization: Algorithms and Techniques* (pp. 16–28). Springer, Berlin, Heidelberg.
20. Bach, E., & Shallit, J. O., (1996). *Algorithmic Number Theory: Efficient Algorithms* (Vol. 1). MIT press.
21. Bach, E., (1990). Number-theoretic algorithms. *Annual Review of Computer Science, 4*(1), 119–172.
22. Bailey, D. H., Lee, K., & Simon, H. D., (1991). Using Strassen's algorithm to accelerate the solution of linear systems. *The Journal of Supercomputing, 4*(4), 357–371.
23. Bakker, M., Riezebos, J., & Teunter, R. H., (2012). Review of inventory systems with deterioration since 2001. *European Journal of Operational Research, 221*(2), 275–284.
24. Baswana, S., Hariharan, R., & Sen, S., (2002). Improved decremental algorithms for maintaining transitive closure and all-pairs shortest

paths. In: *Proceedings of the Thirty-Fourth Annual ACM Symposium on Theory of Computing* (pp. 117–123). ACM.

25. Bateni, M., Hajiaghayi, M., & Zadimoghaddam, M., (2010). Submodular secretary problem and extensions. In: *Approximation, Randomization, and Combinatorial Optimization: Algorithms and Techniques* (pp. 39–52). Springer, Berlin, Heidelberg.
26. Bayer, R., (1972). Symmetric binary B-trees: Data structure and maintenance algorithms. *Acta Informatica, 1*(4), 290–306.
27. Beauchemin, P., Brassard, G., Crépeau, C., Goutier, C., & Pomerance, C., (1988). The generation of random numbers that are probably prime. *Journal of Cryptology, 1*(1), 53–64.
28. Bellare, M., & Sudan, M., (1994). Improved non-approximability results. In: *Proceedings of the Twenty-Sixth Annual ACM Symposium on Theory of Computing* (pp. 184–193). ACM.
29. Bender, M. A., Demaine, E. D., & Farach-Colton, M., (2000). Cache-oblivious B-trees. In: *Foundations of Computer Science, 2000; Proceedings 41st Annual Symposium* (pp. 399–409). IEEE.
30. Ben-Or, M., (1983). Lower bounds for algebraic computation trees. In: *Proceedings of the Fifteenth Annual ACM Symposium on Theory of Computing* (pp. 80–86). ACM.
31. Bent, S. W., & John, J. W., (1985). Finding the median requires 2n comparisons. In: *Proceedings of the Seventeenth Annual ACM Symposium on Theory of Computing* (pp. 213–216). ACM.
32. Bentley, J. L., Haken, D., & Saxe, J. B., (1980). A general method for solving divide-and-conquer recurrences. *ACM SIGACT News, 12*(3), 36–44.
33. Berger, J., & Barkaoui, M., (2004). A parallel hybrid genetic algorithm for the vehicle routing problem with time windows. *Computers & Operations Research, 31*(12), 2037–2053.
34. Berry, M. V., & Howls, C. J., (2012). *Integrals with Coalescing Saddles, 36*, 775–793.
35. Bienstock, D., & McClosky, B., (2012). Tightening simple mixed-integer sets with guaranteed bounds. *Mathematical Programming, 133*(1), 337–363.
36. Bienstock, D., (2008). Approximate formulations for 0-1 knapsack sets. *Operations Research Letters, 36*(3), 317–320.

37. Blasgen, M. W., Casey, R. G., & Eswaran, K. P., (1977). An encoding method for multifield sorting and indexing. *Communications of the ACM, 20*(11), 874–878.
38. Blelloch, G. E., & Gibbons, P. B., (2004). Effectively sharing a cache among threads. In: *Proceedings of the Sixteenth Annual ACM Symposium on Parallelism in Algorithms and Architectures* (pp. 235–244). ACM.
39. Blelloch, G. E., & Greiner, J., (1996). A provable time and space efficient implementation of NESL. In: *ACM SIGPLAN Notices* (Vol. 31, No. 6, pp. 213–225). ACM.
40. Blelloch, G. E., & Maggs, B. M., (1996). Parallel algorithms. *ACM Computing Surveys (CSUR), 28*(1), 51–54.
41. Blelloch, G. E., Hardwick, J. C., Sipelstein, J., Zagha, M., & Chatterjee, S., (1994). Implementation of a portable nested data-parallel language. *Journal of Parallel and Distributed Computing, 21*(1), 4–14.
42. Bloom, N., & Van, R. J., (2002). Patents, real options and firm performance. *The Economic Journal, 112*(478).
43. Blumofe, R. D., & Leiserson, C. E., (1999). Scheduling multithreaded computations by work stealing. *Journal of the ACM (JACM), 46*(5), 720–748.
44. Bollerslev, T., Engle, R. F., & Nelson, D. B., (1994). ARCH models. *Handbook of Econometrics, 4*, 2959–3038.
45. Brassard, G., & Bratley, P., (1996). *Fundamentals of Algorithmics* (Vol. 33). Englewood Cliffs: Prentice Hall.
46. Brent, R. P., (1974). The parallel evaluation of general arithmetic expressions. *Journal of the ACM (JACM), 21*(2), 201–206.
47. Brodnik, A., Miltersen, P. B., & Munro, J. I., (1997). Trans-dichotomous algorithms without multiplication—Some upper and lower bounds. In: *Workshop on Algorithms and Data Structures* (pp. 426–439). Springer, Berlin, Heidelberg.
48. Buchbinder, N., Jain, K., & Singh, M., (2010). Secretary problems via linear programming. In: *International Conference on Integer Programming and Combinatorial Optimization* (pp. 163–176). Springer, Berlin, Heidelberg.
49. Buhler, J. P., Lenstra, H. W., & Pomerance, C., (1993). Factoring integers with the number field sieve. In: *The Development of the Number Field Sieve* (pp. 50–94). Springer, Berlin, Heidelberg.

50. Chan, T. H., (2000). HOMFLY polynomials of some generalized Hopf links. *Journal of Knot Theory and Its Ramifications, 9*(07), 865–883.
51. Chen, L. T., & Davis, L. S., (1990). A parallel algorithm for list ranking image curves in O (log N) time. In: *DARPA Image Understanding Workshop* (pp. 805–815).
52. Cheriyan, J., & Hagerup, T., (1989). A randomized maximum-flow algorithm. In: *Foundations of Computer Science, 30th Annual Symposium* (pp. 118–123). IEEE.
53. Cheriyan, J., & Hagerup, T., (1995). A randomized maximum-flow algorithm. *SIAM Journal on Computing, 24*(2), 203–226.
54. Cheriyan, J., Hagerup, T., & Mehlhorn, K., (1990). Can a maximum flow be computed in o (nm) time?. In: *International Colloquium on Automata, Languages, and Programming* (pp. 235–248). Springer, Berlin, Heidelberg.
55. Cheriyan, J., Hagerup, T., & Mehlhorn, K., (1996). An o(n^3)-time maximum-flow algorithm. *SIAM Journal on Computing, 25*(6), 1144–1170.
56. Ciurea, E., & Ciupala, L., (2001). Algorithms for minimum flows. *Computer Science Journal of Moldova, 9*(3), 27.
57. Conforti, M., Wolsey, L. A., & Zambelli, G., (2010). Projecting an extended formulation for mixed-integer covers on bipartite graphs. *Mathematics of Operations Research, 35*(3), 603–623.
58. Courcoubetis, C., Vardi, M., Wolper, P., & Yannakakis, M., (1992). Memory-efficient algorithms for the verification of temporal properties. *Formal Methods in System Design, 1*(2, 3), 275–288.
59. D'Aristotile, A., Diaconis, P., & Newman, C. M., (2003). Brownian motion and the classical groups. *Lecture Notes-Monograph Series*, 97–116.
60. Damgård, I., Landrock, P., & Pomerance, C., (1993). Average case error estimates for the strong probable prime test. *Mathematics of Computation, 61*(203), 177–194.
61. Dehling, H., & Philipp, W., (2002). Empirical process techniques for dependent data. In: *Empirical Process Techniques for Dependent Data* (pp. 3–113). Birkhäuser, Boston, MA.
62. Dengiz, B., Altiparmak, F., & Smith, A. E., (1997). Efficient optimization of all-terminal reliable networks, using an evolutionary approach. *IEEE Transactions on Reliability, 46*(1), 18–26.

63. Dickerson, M., Eppstein, D., Goodrich, M. T., & Meng, J. Y., (2003). Confluent drawings: Visualizing non-planar diagrams in a planar way. In: *International Symposium on Graph Drawing* (pp. 1–12). Springer, Berlin, Heidelberg.

64. Didier, F., (2009). *Efficient Erasure Decoding of Reed-Solomon Codes* (Vol. 1, pp. 1–22). arXiv preprint arXiv:0901.1886.

65. Drmota, M., & Szpankowski, W., (2013). A master theorem for discrete divide and conquer recurrences. *Journal of the ACM (JACM), 60*(3), 16.

66. Du, W., & Atallah, M. J., (2001). Protocols for secure remote database access with approximate matching. In: *E-Commerce Security and Privacy* (pp. 87–111). Springer, Boston, MA.

67. Duda, R. O., Hart, P. E., & Stork, D. G., (1973). *Pattern Classification* (Vol. 2). New York: Wiley.

68. Eppstein, D., Goodrich, M. T., & Sun, J. Z., (2005). The skip quadtree: A simple dynamic data structure for multidimensional data. In: *Proceedings of the Twenty-First Annual Symposium on Computational Geometry* (pp. 296–305). ACM.

69. Eppstein, D., Goodrich, M. T., & Sun, J. Z., (2008). Skip quadtrees: Dynamic data structures for multidimensional point sets. *International Journal of Computational Geometry & Applications, 18*(01n02), 131–160.

70. Faenza, Y., & Sanità, L., (2015). On the existence of compact ε-approximated formulations for knapsack in the original space. *Operations Research Letters, 43*(3), 339–342.

71. Festa, P., & Resende, M. G., (2002). GRASP: An annotated bibliography. In: *Essays and Surveys in Metaheuristics* (Vol. 1, pp. 325–367). Springer, Boston, MA.

72. Fourier, J. B. J., (1890). In: Darboux, G., (ed.), *From 1824, Republished as Second Extrait in Oeuvres de Fourier, Tome II* (pp. 38–42). Gauthier-Villars, Paris.

73. Fourier, J. B. J., (1973). In: Tome II, G. D., (ed.), *From 1824, Republished as Second Extrait in Oeuvres de Fourier*. Gauthier-Villars, Paris 1890, see DA Kohler, Translation of a report by Fourier on his work on linear inequalities.

74. Fredman, M. L., & Willard, D. E., (1993). Surpassing the information theoretic bound with fusion trees. *Journal of Computer and System Sciences, 47*(3), 424–436.

75. Friedl, K., & Sudan, M., (1995). Some improvements to total degree tests. In: *Theory of Computing and Systems, Proceedings Third Israel Symposium* (pp. 190–198). IEEE.

76. Frigo, M., Leiserson, C. E., & Randall, K. H., (1998). The implementation of the Cilk-5 multithreaded language. *ACM SIGPLAN Notices, 33*(5), 212–223.

77. Fussenegger, F., & Gabow, H. N., (1979). A counting approach to lower bounds for selection problems. *Journal of the ACM (JACM), 26*(2), 227–238.

78. Glasser, K. S., & Austin, B. R. H., (1983). The d choice secretary problem. *Sequential Analysis, 2*(3), 177–199.

79. Goodrich, M. T., Atallah, M. J., & Tamassia, R., (2005). Indexing information for data forensics. In: *International Conference on Applied Cryptography and Network Security* (pp. 206–221). Springer, Berlin, Heidelberg.

80. Graefe, G., (2011). Modern B-tree techniques. *Foundations and Trends® in Databases, 3*(4), 203–402.

81. Hall, M., Frank, E., Holmes, G., Pfahringer, Badel'son-Vel'skii, G. M., & Landis, E. M., (1962). *Partial Differential Equations of Elliptic Type* (Vol. 1, pp. 1–22). Springer, Berlin.

82. Herr, D. G., (1980). On the history of the use of geometry in the general linear model. *The American Statistician, 34*(1), 43–47.

83. Icking, C., Klein, R., & Ottmann, T., (1987). Priority search trees in secondary memory. In: *International Workshop on Graph-Theoretic Concepts in Computer Science* (pp. 84–93). Springer, Berlin, Heidelberg.

84. IGARASHI, Y., SADO, K., & SAGA, K., (1987). Fast parallel sorts on a practical sized mesh-connected processor array. *IEICE Transactions (1976–1990), 70*(1), 56–64.

85. John, J. W., (1988). A new lower bound for the set-partitioning problem. *SIAM Journal on Computing, 17*(4), 640–647.

86. Kim, S. H., & Pomerance, C., (1989). The probability that a random probable prime is composite. *Mathematics of Computation, 53*(188), 721–741.

87. King, D. J., (1995). Functional binomial queues. In: *Functional Programming, Glasgow 1994* (Vol. 1, pp. 141–150). Springer, London.
88. Kirkpatrick, D. G., (1981). A unified lower bound for selection and set partitioning problems. *Journal of the ACM (JACM), 28*(1), 150–165.
89. Kwong, Y. S., & Wood, D., (1982). A new method for concurrency in B-trees. *IEEE Transactions on Software Engineering,* (3), 211–222.
90. Leighton, T., (1996). *Notes on Better Master Theorems for Divide-and-Conquer Recurrences.* Manuscript. Massachusetts Institute of Technology.
91. Maurer, S. B., (1985). The lessons of Williamstown. In: *New Directions in Two-Year College Mathematics* (Vol. 1, pp. 255–270). Springer, New York, NY.
92. Meijer, H., & Akl, S. G., (1987). Optimal computation of prefix sums on a binary tree of processors. *International Journal of Parallel Programming, 16*(2), 127–136.
93. Meijer, H., & Akl, S. G., (1988). Bit serial addition trees and their applications. *Computing, 40*(1), 9–17.
94. Monier, L., (1980). Evaluation and comparison of two efficient probabilistic primality testing algorithms. *Theoretical Computer Science, 12*(1), 97–108.
95. Morain, F., (1988). *Implementation of the Atkin-Goldwasser-Kilian Primality Testing Algorithm*. Doctoral dissertation, INRIA.
96. Munro, J. I., & Poblete, P. V., (1982). *A Lower Bound for Determining the Median.* Faculty of Mathematics, University of Waterloo.
97. Nelson, D. B., & Foster, D. P., (1992). *Filtering and Forecasting with Mis-Specified ARCH Models II: Making the Right Forecast with the Wrong Model.*(pp.1-25)
98. Nievergelt, J., (1974). Binary search trees and file organization. *ACM Computing Surveys (CSUR), 6*(3), 195–207.
99. Park, T. G., & Oldfield, J. V., (1993). Minimum spanning tree generation with content-addressable memory. *Electronics Letters, 29*(11), 1037–1039.
100. Phillips, S., & Westbrook, J., (1993). Online load balancing and network flow. In: *Proceedings of the Twenty-Fifth Annual ACM Symposium on Theory of Computing* (pp. 402–411). ACM.

101. Polishchuk, A., & Spielman, D. A., (1994). Nearly-linear size holographic proofs. In: *Proceedings of the Twenty-Sixth Annual ACM symposium on Theory of Computing* (pp. 194–203). ACM.

102. Prechelt, L., (1993). *Measurements of MasPar MP-1216A Communication Operations*. Univ., Fak. Für Informatik.

103. Price, G. B., (1973). Telescoping sums and the summation of sequences. *The Two-Year College Mathematics Journal, 4*(2), 16–29.

104. Ramadge, P. J., & Wonham, W. M., (1989). The control of discrete event systems. *Proceedings of the IEEE, 77*(1), 81–98.

105. Raman, R., (1996). Priority queues: Small, monotone and trans-dichotomous. In: *European Symposium on Algorithms* (pp. 121–137). Springer, Berlin, Heidelberg.

106. Regli, W. C., (1992). *A Survey of Automated Feature Recognition Techniques, 1*, 1–22.

107. Roura, S., (2001). Improved master theorems for divide-and-conquer recurrences. *Journal of the ACM (JACM), 48*(2), 170–205.

108. Sardelis, D. A., & Valahas, T. M., (1999). Decision making: A golden rule. *The American Mathematical Monthly, 106*(3), 215–226.

109. Schönhage, A., Paterson, M., & Pippenger, N., (1976). Finding the median. *Journal of Computer and System Sciences, 13*(2), 184–199.

110. Shen, Z., & Marston, C. M., (1995). A study of a dice problem. *Applied Mathematics and Computation, 73*(2, 3), 231–247.

111. Smith, R. S., (1986). Rolle over Lagrange—Another shot at the mean value theorem. *The College Mathematics Journal, 17*(5), 403–406.

112. Snyder, L., (1984). *Parallel Programming and the Poker Programming Environment* (No. TR-84-04-02, pp. 1–30). Washington Univ. Seattle Dept. of Computer Science.

113. Stock, J. H., & Watson, M. W., (2001). Vector autoregressions. *Journal of Economic Perspectives, 15*(4), 101–115.

114. Sudan, M., (1992). Efficient checking of polynomials and proofs and the hardness of approximation problems. *Lecture Notes in Computer Science,* 1001.

115. Szymanski, T. G., & Van, W. C. J., (1983). Space efficient algorithms for VLSI artwork analysis. In: *Proceedings of the 20th Design Automation Conference* (pp. 734–739). IEEE Press.

116. Szymanski, T. G., (1975). *A Special Case of the Maximal Common Subsequence Problem*. Technical Report TR-170, Computer Science Laboratory, Princeton University.

117. Thorup, M., (1997). *Faster Deterministic Sorting and Priority Queues in Linear Space* (pp. 550–555). Max-Planck-Institut für Informatik.

118. Vanderbei, R. J., (1980). The optimal choice of a subset of a population. *Mathematics of Operations Research, 5*(4), 481–486.

119. Verma, R. M., (1994). A general method and a master theorem for divide-and-conquer recurrences with applications. *Journal of Algorithms, 16*(1), 67–79.

120. Verma, R. M., (1997). General techniques for analyzing recursive algorithms with applications. *SIAM Journal on Computing, 26*(2), 568–581.

121. Wallace, L., Keil, M., & Rai, A., (2004). Understanding software project risk: A cluster analysis. *Information & Management, 42*(1), 115–125.

122. Wang, Y., (2008). Topology control for wireless sensor networks. In: *Wireless Sensor Networks and Applications* (Vol. 1, pp. 113–147). Springer, Boston, MA.

123. Wilf, H. S., (1984). A bijection in the theory of derangements. *Mathematics Magazine, 57*(1), 37–40.

124. Williamson, J., (2002). Probability logic. In: *Studies in Logic and Practical Reasoning* (Vol. 1, pp. 397–424). Elsevier.

125. Wilson, J. G., (1991). Optimal choice and assignment of the best m of n randomly arriving items. *Stochastic Processes and Their Applications, 39*(2), 325–343.

126. Winograd, S., (1970). On the algebraic complexity of functions. In: *Actes du Congres International des Mathématiciens* (Vol. 3, pp. 283–288).

127. Wunderlich, M. C., (1983). A performance analysis of a simple prime-testing algorithm. *Mathematics of Computation, 40*(162), 709–714.

128. Xiaodong, W., & Qingxiang, F., (1996). A frame for general divide-and-conquer recurrences. *Information Processing Letters, 59*(1), 45–51.

129. Yap, C., (2011). A real elementary approach to the master recurrence and generalizations. In: *International Conference on Theory and Applications of Models of Computation* (pp. 14–26). Springer, Berlin, Heidelberg.

130. Zhu, X., & Wilhelm, W. E., (2006). Scheduling and lot sizing with sequence-dependent setup: A literature review. *IIE Transactions, 38*(11), 987–1007.

CHAPTER 2

CLASSIFICATION OF ALGORITHMS

CONTENTS

2.1. INTRODUCTION

An algorithm is a technique or group of methods for completing a problem-solving activity. The term algorithm, which comes from Medieval Latin, refers to more than only computer programming. There are many different sorts of algorithms for various issues. The belief that there are only a finite number of algorithms and that one must learn all of them is a fallacy that leads many potential programers to turn to lawn maintenance as a source of revenue to cover their expenses (Rabin, 1977; Cook, 1983, 1987).

When an issue develops throughout the course of developing the complete software, an algorithm is created. A variety of classification schemes for algorithms are discussed in detail in this chapter. There is no single "correct" classification that applies to all situations. When comparing the tasks of classifying algorithms and attributing them, it is important to remember that the former is more difficult (Karp, 1986; Virrer and Simons, 1986). Following a discussion of the labeling procedure for a given algorithm (for example, the divide-and-conquer algorithm), we will examine the various methods for examining algorithms in greater depth. Frequently, the labels with which the algorithms are classified are quite useful and assist in the selection of the most appropriate type of analysis to perform (Figure 2.1) (Cormen and Leiserson, 1989; Stockmeyer and Meyer, 2002).

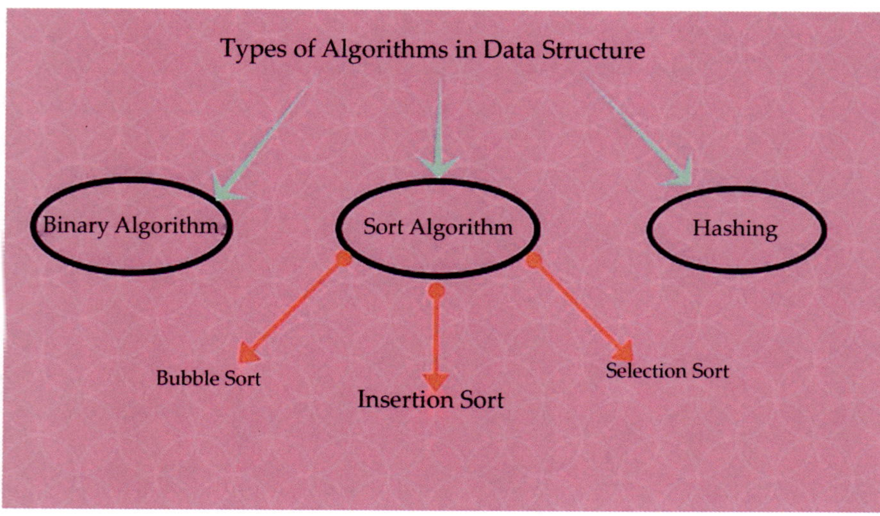

Figure 2.1. Major types of data structure algorithms.

Source: http://www.codekul.com/blog/types-algorithms-data-structures-every-programmer-know/.

The number of fundamental jobs that an algorithm completes is used to determine its speed or efficiency. Consider the case of an algorithm whose input is N and which executes a large number of operations. The characteristics of an algorithm are defined by the connection between the number of tasks completed and the time required to complete each task (Dayde, 1996; Panda et al., 1997; D'Alberto, 2000). Therefore, it must be noted that every algorithm relates to a certain class of algorithms. In the increasing order of their growth, algorithms are classified as:

- Constant time algorithm;
- Logarithmic algorithm;
- Linear time algorithm;
- Polynomial-time algorithm;
- Exponential time algorithm.

2.2. DETERMINISTIC AND RANDOMIZED ALGORITHMS

The following is an example of one of the most significant (and unique) differences that may be used to determine if a particular algorithm is deterministic or randomized (Codenotti and Simon, 1998; Kgström and Van Loan, 1998).

On a given input, Deterministic algorithms produce the same outcomes by using the same calculation steps, whereas Randomized algorithms, in contrast to Deterministic algorithms, throw coins during the implementation process. It is possible that the sequence in which the algorithm is implemented or the conclusion of the algorithm will change for each try on a specific input (Flajolet et al., 1990; Nisan and Wigderson, 1995).

For randomized algorithms, there are two more subcategories.

- *Monte Carlo* algorithms; and
- *Las Vegas* algorithms.

A Las Vegas algorithm will always provide the same output for a given input. Randomization will have no effect on the order of the core implementations.

In the case of Monte Carlo algorithms, the output of these algorithms may vary, if not be incorrect. A Monte Carlo method, on the other hand, produces the right result with a high degree of certainty (Garey, 1979; Kukuk, 1997; Garey and Johnson, 2002).

The issue that arises at this point is: For what purpose are randomized algorithms being developed? The computation/processing may change based on the outcome of the coin toss (Hirschberg and Wong, 1976; Kannan, 1980). However, despite the fact that Monte Carlo algorithms do not produce precise results, they are still sought after for the following reasons:

- Randomized algorithms typically have the effect of troubling the input, which is why they are used. To put it another way, the input appears to be random, and as a result, undesirable situations are rarely seen.
- When it comes to conceptual implementation, randomized algorithms are typically quite stress-free. When compared to their deterministic equivalents, they are typically significantly superior in terms of runtime performance (Figure 2.2).

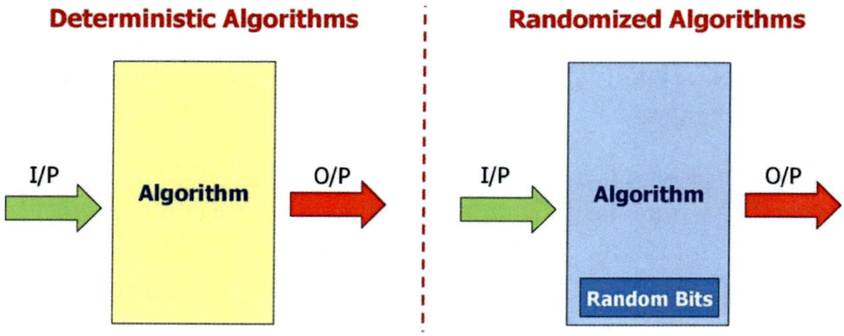

Figure 2.2. Illustration of deterministic and randomized algorithms.

Source: https://slideplayer.com/slide/17834668/.

2.3. ONLINE VS. OFFLINE ALGORITHMS

One of the main differences in determining whether a particular algorithm is online or offline is mentioned below.

Online Algorithms are algorithms in which the inputs are unknown at the start. The inputs to algorithms are usually known ahead of time. However, in the case of Online, they are given to them (Figure 2.3) (Chazelle et al., 1989; Roucairol, 1996; Goodman and O'Rourke, 1997).

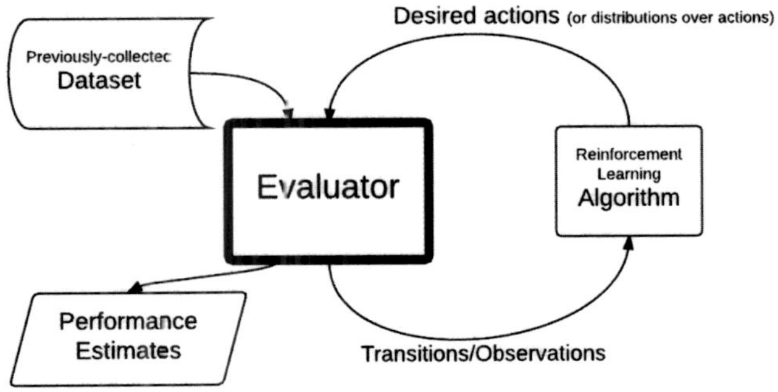

Figure 2.3. Offline evaluation of online reinforcement learning algorithms.

Source: https://grail.cs.washington.edu/projects/nonstationaryeval/.

Even though it appears to be a minor issue, its implications for the design and analysis of the algorithms are significant. The competitiveness of these algorithms is typically tested by applying competitiveness knowledge to them. In the worst-case scenario, this examination takes the most time, and when compared to the best algorithm, it normally takes the most time. A ski problem is an example of a problem that may be found on the internet (Wilson and Pawley, 1988; Xiang et al., 2004; Liu et al., 2007).

Every day that a skier goes skiing, he or she must determine whether to purchase or rent skis, at least until the point at which he or she decides to purchase skis is reached. Because of the unpredictable weather, it is impossible to predict how many days a skier will be able to enjoy the sport. Assume that T is the number of days he or she will be skiing. B represents the cost of purchasing skis; 1 unit represents the cost of renting skis (Wierzbicki et al., 2002; Li, 2004; Pereira, 2009).

2.4. EXACT, APPROXIMATE, HEURISTIC, AND OPERATIONAL ALGORITHMS

The majority of algorithms are designed with optimization in mind, such as calculating the direct path, alignment, or marginal edit distance (Aydin and Fogarty, 2004). When a goal is stated, an exact algorithm concentrates on computing the best answer. This is relatively costly in terms of runtime or memory, and it is not practicable for huge inputs (Tran et al., 2004; Thallner and Moser, 2005; Xie et al., 2006).

Other approaches are tested at this time. Approximation algorithms are one of the strategies that focus on computing a solution that is just a firm, defined factor poorer than the optimum answer (Ratakonda and Turaga, 2008). This indicates that an algorithm is a c approximation if it can guarantee that the answer it generates is never worse than the factor c when compared to the best solution (Xiang et al., 2003; Aydin and Yigit, 2005).

Heuristic algorithms, on the other hand, attempt to provide the best answer without guaranteeing that it will always be the best solution. It is frequently simple to construct a counterexample. A good heuristics algorithm is always at or near the optimal value (Restrepo et al., 2004; Sevkli and Aydin, 2006).

Finally, there are algorithms that do not prioritize the optimization of objective functions (Yigit et al., 2004; Sevkli and Guner, 2006). Because they tie a succession of computing tasks controlled by an expert but not in aggregate with a specific goal function, these algorithms are referred to be operational (e.g., ClustalW).

Consider the Traveling Salesman Problem, which has triangle inequality for n cities. This is an example of a problem that is NP-hard. A greedy, deterministic technique is described below that generates two approximations for the traveling salesman problem—the triangle inequality produced in time $O(n^2)$:

- A minimal spanning tree (T) is computed for the whole graph bounded by n cities;
- All edges of the spanning tree *T are duplicated by* producing a Eulerian graph T'. The Eulerian path is then found in T';
- By using clever shortcuts, the Eulerian cycle is transformed into the Hamiltonian cycle.

2.5. CLASSIFICATION ACCORDING TO THE MAIN CONCEPT

The algorithms may be classified as follows using the primary algorithmic paradigm that we are familiar with:

- Simple recursive algorithm;
- Divide-and-conquer algorithm;
- Dynamic programming algorithm;
- Backtracking algorithm;
- Greedy algorithm;
- Brute force algorithm;
- Branch-and-bound algorithm.

2.5.1. Simple Recursive Algorithm

This kind of algorithm has the following characteristics (Figure 2.4):

- It fixes the fundamental issues right away.
- It continues to work on the less difficult subproblem.
- It goes through some extra steps to convert the answer to the easier subproblem into the solution to the given problem. Examples are discussed below:
- Counting how many essentials are in a list:
- If the specified list is empty, return zero; otherwise;
- Remove the first item from the equation and add up the remaining requirements in the list;
- To the result, add one.
- Checking if a value is present in a list:
- If the supplied list is empty, return false, otherwise;
- If the first object in the supplied list is the requested value, return true; otherwise;
- After excluding the first object, see if the value appears elsewhere in the list.

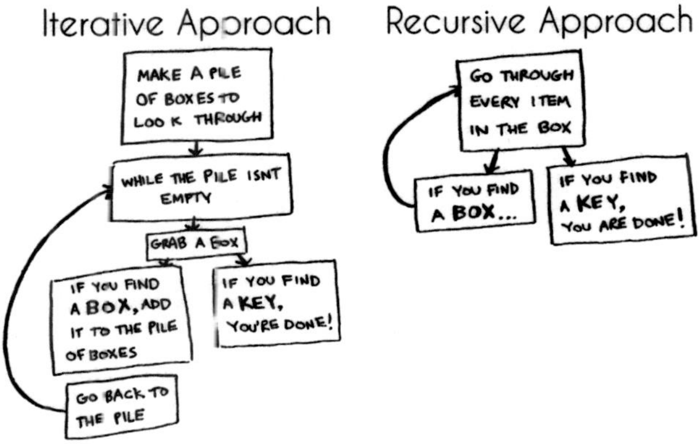

Figure 2.4. Iterative and recursive approaches are compared.

Source: https://www.freecodecamp.org/news/how-recursion-works-explained-with-flowcharts-and-a-video-de61f40cb7f9/.

2.5.2. Divide-and-Conquer Algorithm

The size of a task is divided by a predetermined factor in this approach. Only a small percentage of the genuine problem is treated in each cycle. This category includes a few of the most efficient and effective algorithms. This type of algorithm has a logarithmic runtime (Battiti and Tecchiolli, 1994; Yigit et al., 2006). This algorithm comprises two parts.

- The original problem is broken down into smaller, related subproblems, which are then solved in a recursive manner; and
- The subproblems' solutions are merged to provide the solution to the original problem.

Traditionally, a divide-and-conquer algorithm is defined as one that comprises two or more recursive calls. Consider the following two case studies:

- Quicksort:
  - Divide the array into two equal pieces, then sort each component quickly;
  - Combining the partitioned portions requires no additional effort.
- Merge sort:
  - After slicing the array in half, merge sort each part;
  - By combining the two organized arrays, you may create a single organized array.

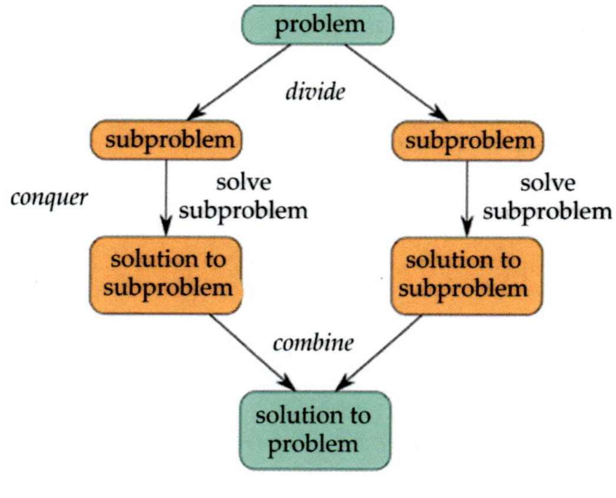

Source: https://www.khanacademy.org/computing/computer-science/algorithms/merge-sort/a/divide-and-conquer-algorithms.

2.5.3. Dynamic Programming Algorithm

The term 'dynamic' refers to the manner through which the algorithm computes the outcome of the computation. A solution to one problem may be contingent on the solution of one or more sub-problems in another situation (Taillard, 1991, 1995). It exemplifies the characteristic of coinciding sub-problems in a straightforward manner. As a result, it is possible that we will have to recalculate the same numbers for sub-problems over and over to answer the main problem. As a result, counting cycles is pointless (Fleurent and Ferland, 1994).

A technique or procedure called dynamic programming can be used to alleviate the frustration caused by these useless computer cycles. As a result of this method, the conclusion of each sub-problem is memorized and is utilized the outcome anytime is required, rather than having to recalculate the result again and over (Glover, 1989; Skorin-Kapov, 1990).

In this case, space is exchanged for time. As an example, more space is employed to grab the computed numbers, allowing the execution speed to be greatly boosted. The connection for the Nth Fibonacci number is the greatest example of a problem that involves several sub-problems that are related to each other. The Fibonacci number is expressed by the equation $F(n) = F(n - 1) + F(n - 2)$.

The statement above demonstrates that the Nth Fibonacci number is reliant on the two numbers that came before it. To compute F(n) in a predictable manner, the computations must be carried out in the manner outlined below. The colored values that are similar in appearance are those that will be computed repeatedly. Take note that $F(n - 2)$ is computed twice, $F(n - 3)$ is computed three times, and so on… As a result, we are wasting a significant amount of time. As it turns out, this recursion will do $$2N$$ operations for every given N, and it is completely insoluble on a current PC for any N more than 40 within one to two years on the most recent generation of computers (Figure 2.5) (Shi, 2001; Hu et al., 2003).

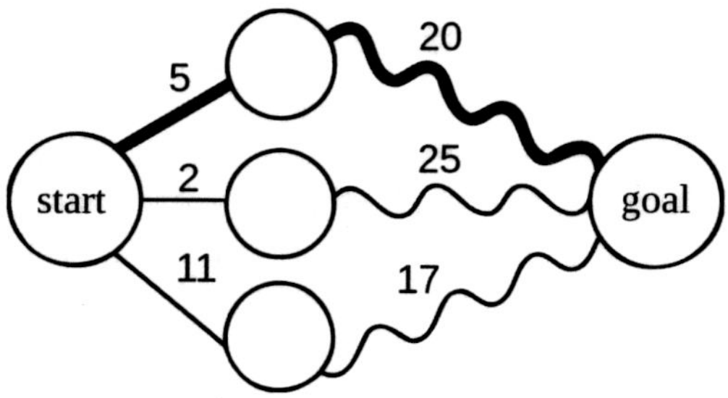

Figure 2.5. The dynamic programming algorithm is depicted in a diagram.

Source: https://en.wikipedia.org/wiki/Dynamic_programming.

The greatest potential answer to this problem is to save every value when computing it and retrieve it instead of having to compute it all over again. Because of this, the exponential time method is changed into the linear time algorithm and vice versa (Kennedy et al., 2001). The use of dynamic programming techniques is extremely important to accelerate the solutions to issues that have concurrent subproblems (Kennedy and Mendes, 2002, 2006). This dynamic algorithm memorizes the previous outcomes and then makes use of them to find the new future outcomes (He et al., 2004). The dynamic programming approach is typically used to solve optimization issues in situations where:

- The optimal option must be discovered amid a plethora of other possibilities;
- Coinciding sub-problem and optimum substructure are needed;
- Optimum substructure: an Optimum solution that is composed of optimum solutions to the constituent issues;
- Coinciding sub-problems: Bottom-up methods allow you to preserve and reuse the solutions you find for sub-problems you encounter.

This method differs from the Divide-and-Conquer method, in which the sub-problems seldom overlap. In bioinformatics, there are several instances of this method. For example:

- Obtaining the best pairwise alignment possible;
  - **Optimum Substructure:** The answers for the most optimal designs of smaller prefixes are incorporated into the arrangement of two prefixes as well.

- **Coinciding Sub-Problems:** The stored results of the design of three sub-problems are used to get the best arrangement of two prefixes.

- In an HMM, we are calculating a Viterbi path:
  - **Optimum Substructure:** A Viterbi route for an input prefix that ends in a condition of HMM is composed of shorter Viterbi routes for minor sections of the input and the other HMM conditions, as shown in the figure 2.5 .
  - **Coinciding Sub-Problems:** To find a solution for a Viterbi route for input prefix that ends in a condition of HMM, the saved outcomes of the Viterbi pathways for smaller input prefix and the HMM requirements must be used in conjunction with the HMM conditions.

2.5.4. Backtracking Algorithm

This method is quite like the Brute Force algorithm mentioned later in this chapter. There are significant differences between the two algorithms. In the brute force technique, each possible solution combination is created and checked to see if it is legitimate (Kröse et al., 1993; Pham and Karaboga, 2012). In contrast to the backtracking method, every time a solution is created, it is tested, and only if it meets all of the requirements are further solutions generated; otherwise, the approach is backtracked, and an alternative path for finding the answer is taken (Wu et al., 2005). The N Queens issue is a well-known example of this type of problem. An N × N chessboard is given, according to the N Queens. On the chessboard, N queens are arranged in such a way that none of them are attacked by the other queens (Figure 2.6).

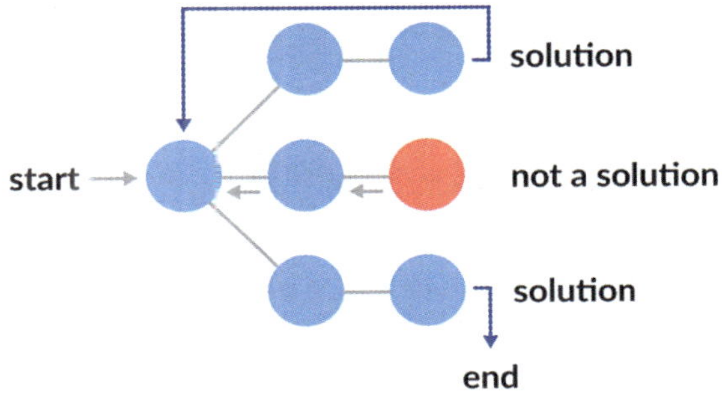

Figure 2.6. Backtracking algorithm representation.

Source: https://www.programiz.com/dsa/backtracking-algorithm.

This is accomplished by shifting a queen in each column and appropriate row. The status of a queen is verified every time she is moved to ensure that she is not under assault. If this is the case, a different cell inside that column is picked to be the location of the queen. You may think of the process in terms of a tree. Every node in the tree represents a chessboard, each with a unique configuration. If we are unable to move at any point, we can retrace our steps back to the previous node and go on by increasing the number of other nodes.

The advantage of this strategy over the Brute force algorithm is that it generates a smaller number of candidates when compared to the Brute force algorithm. With the aid of this strategy, it is possible to isolate the viable answers in a relatively short period of time and with great effectiveness.

Consider the case of an eight × eight chessboard; if the Brute Force technique is used, then 4,426,165,368 solutions must be developed, and each one must be checked before a solution is accepted. The quantity of solutions is decreased to around 40,320 in the manner that has been explained.

There are several advantages to using the depth-first recursive hunt outlined above. The following are some of these advantages:

- Tests to determine whether the solution has been constructed, and if it has, it returns the solution; otherwise.
- For each option that can be created at this time, a separate document is required:
  - Make your choice;
  - Recur;
  - If the recursion produces a result, return it.
- Return failure if there is no other option.

Consider drawing or painting a map with no more than four colors, i.e.

Color (Country *n*)

- If all of the countries are colored ($n >$ no of countries) return success; otherwise.
- For each of the four hues, there is a c.
- If the country isn't close to one of the colored countries:
  - Color c is used to color country n;
  - Color country n + 1 in a recursive manner;
  - Return success if it was successful.

- Return failure (if the loop still exits).

2.5.5. Greedy Algorithm

This approach delivers superior results for optimization issues on occasion, and it typically works in phases. At every step:

- You can attain the finest results possible right now without having to worry about future outcomes;
- One can attain a global ideal by picking a local optimal at each stage along the way.

In my issues, using this approach results in the best answer. This approach works well with optimization issues. We will develop a locally optimal solution at each stage of this method, which will lead to a globally optimal solution. We can't back out of a decision once we've made it. Verifying the validity of this method is critical since not all greedy algorithms produce the best answer worldwide (Creput et al., 2005). Consider the scenario in which you are given a specified quantity of coins and asked to make a specific amount of money with those coins (Figure 2.7).

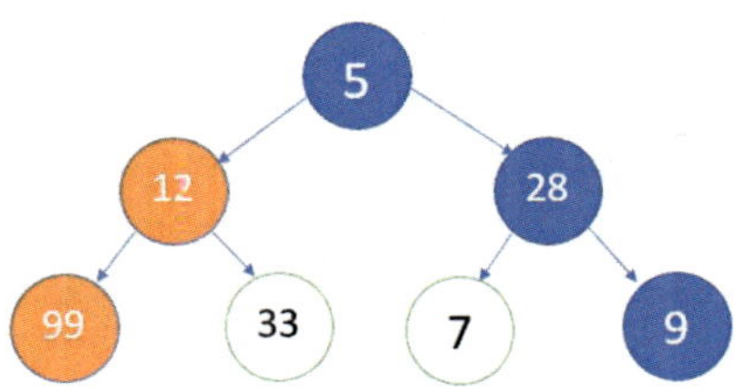

Figure 2.7. A greedy algorithm's numerical representation.

Source: https://www.techopedia.com/definition/16931/greedy-algorithm.

This method is frequently quite successful and always yields the best solution for certain types of situations. Let's look at another scenario. Let's say a guy wishes to count out a specified amount of money using the fewest possible notes and coins. Here, the greedy algorithm will choose the note or coin with the highest potential value that does not exceed a certain threshold. For example, we have a number of options for earning $6.39 dollars:

- A bill for $5;
- A one-dollar note, for a total of six dollars;
- 6.25 dollars with a 25-cent coin;

- To make 6.35 dollars, you'll need a ten-cent coin;
- To make $6.39, you'll need four one-cent coins.

The greedy algorithm always finds the best answer for the number of dollars.

2.5.6. Brute Force Algorithm

Using the fundamental specification, this method solves the issue in a straightforward manner. Although this approach is often the simplest to implement, there are several disadvantages to using it to solve a problem. This can only be used to solve issues with small input sizes that are often quite sluggish (Divina and Marchiori, 2005). This algorithm explores all options in search of a good solution to the problem. Brute force algorithm can be categorized as (Figure 2.8):

- **Optimizing:** This method assists us in identifying the most optimal feasible option. This may involve discovering all of the solutions if the value of the most wonderful option is already known; nevertheless, after the finest potential answer has been identified, it may be necessary to halt searching for other alternatives.
- **Satisficing:** After finding the best potential solution to the problem, this algorithm comes to a complete stop.

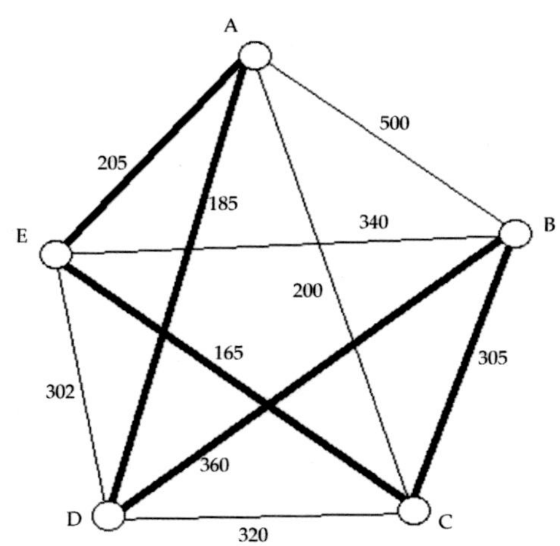

Figure 2.8. The brute force algorithm is depicted in graphical form.

Source: http://www.people.vcu.edu/~gasmerom/MAT131/brutefrc.html.

2.5.7. Branch-and-Bound Algorithm

This type of method is typically employed to assist in the solution of situations when optimization is necessary. An initial tree comprising all the subproblems is formed as soon as the algorithm begins its calculations. The initial problem for which the algorithm is being built is referred to as the root problem. When solving a certain issue, a specific procedure is used to generate the lower and upper jumps (Gunn, 1998; Vapnik, 2013). Each node has its own bonding mechanism, which is implemented.

- If the bounds are comparable, it is possible to estimate a feasible solution to the specific subproblem;
- The issue associated with that specific node should be divided into two subproblems, which should be formed inside the children's nodes if the boundaries are not comparable.

Continue trimming the portions of a tree using the best solution available until all the nodes have been cut (Figure 2.9).

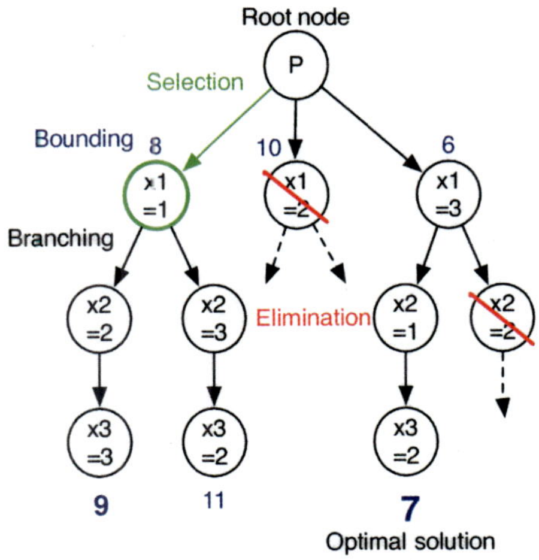

Figure 2.9. The sequential branch-and-bound method is depicted in this diagram.

Source: https://www.researchgate.net/figure/Illustration-of-the-sequential-branch-and-bound-algorithm_fig1_281015427.

The following is an example of the above-described method in the case of the traveling salesman issue:

- The salesperson must visit all n cities at least once, and he wishes to reduce the overall distance traveled.
- The main issue now is to discover the quickest path through all n cities while visiting each one at least once.
- Split the node into two more child issues:
  - The quickest way to see the city for a start; and
  - The shortest route to avoid visiting the city for a start.
- Continue to subdivide the tree in the same manner as it grows.

REFERENCES

1. Aydin, M. E., & Fogarty, T. C., (2004). A distributed evolutionary simulated annealing algorithm for combinatorial optimization problems. *Journal of Heuristics, 10*(3), 269–292.
2. Aydin, M. E., & Fogarty, T. C., (2004). A simulated annealing algorithm for multi-agent systems: A job-shop scheduling application. *Journal of Intelligent Manufacturing, 15*(6), 805–814.
3. Aydin, M. E., & Yigit, V., (2005). 12 parallel simulated annealing. *Parallel Metaheuristics: A New Class of Algorithms, 47*, 267.
4. Battiti, R., & Tecchiolli, G., (1994). The reactive tabu search. *ORSA Journal on Computing, 6*(2), 126–140.
5. Chazelle, B., Edelsbrunner, H., Guibas, L., & Sharir, M., (1989). Lines in space-combinators, algorithms and applications. In: *Proceedings of the Twenty-First Annual ACM Symposium on Theory of Computing* (pp. 382–393). ACM.
6. Codenotti, B., & Simon, J., (1998). *On the Complexity of the Discrete Fourier Transform and Related Linear Transforms Preliminary Version (Vol.1, pp.1-10)*.
7. Cook, S. A., (1983). An overview of computational complexity. *Communications of the ACM, 26*(6), 400–408.
8. Cook, S. A., (1987). Prehl'ad theory of complexity. Pokroky Matematiky, Fyziky a Astronomie, *32*(1), 12–29.
9. Cormen, T. H., Leiserson, C. E., & Rivest, R. L., (1989). *Introduction to Algorithms*. MIT Press. Cambridge, Massachusetts London, England.
10. Creput, J. C., Koukam, A., Lissajoux, T., & Caminada, A., (2005). Automatic mesh generation for mobile network dimensioning using evolutionary approach. *IEEE Transactions on Evolutionary Computation, 9*(1), 18–30.
11. D'Alberto, P., (2000). *Performance Evaluation of Data Locality Exploitation*. University of Bologna, Dept. of Computer Science, Tech. Rep.
12. Dayde, M. J., & Dayde, I. S., (1996). *A Blocked Implementation of Level 3 BLAS for RISC Processors*. TR_PA_96_06. Available on line https://link.springer.com/chapter/10.1007/3-540-44688-5_3
13. Divina, F., & Marchiori, E., (2005). Handling continuous attributes in an evolutionary inductive learner. *IEEE Transactions on Evolutionary Computation, 9*(1), 31–43.

14. Eberhart, R. C., & Shi, Y., (2001). Tracking and optimizing dynamic systems with particle swarms. In: *Evolutionary Computation, 2001; Proceedings of the 2001 Congress* (Vol. 1, pp. 94–100). IEEE.
15. Eberhart, R. C., & Shi, Y., (2004). Guest editorial special issue on particle swarm optimization. *IEEE Transactions on Evolutionary Computation, 8*(3), 201–203.
16. Eberhart, R. C., Shi, Y., & Kennedy, J., (2001). *Swarm Intelligence*. Elsevier.
17. Eberhart, R. C., Shi, Y., & Kennedy, J., (2001). *Swarm Intelligence (The Morgan Kaufmann Series in Evolutionary Computation). (Vol. 1, pp.5-7)*
18. Flajolet, P., Puech, C., Robson, J. M., & Gonnet, G., (1990). *The Analysis of Multidimensional Searching in Quad-Trees*. Doctoral dissertation, INRIA.
19. Fleurent, C., & Ferland, J. A., (1994). Genetic hybrids for the quadratic assignment problem. *Quadratic Assignment and Related Problems, 16*, 173–187.
20. Garey, M. R., & Johnson, D. S., (2002). *Computers and Intractability* (Vol. 29). New York: WH freeman.
21. Garey, M. R., (1979). In: Johnson, D. S., (ed.), *Computers and Intractability: A Guide to the Theory of NP-Completeness*.(pp.2-4)
22. Glover, F., (1989). Tabu search—Part I. *ORSA Journal on Computing, 1*(3), 190–206.
23. Goodman, J. E., & O'Rourke, J., (1997). *Handbook of Discrete and Computational Geometry* (Vol. 6). CRC Press series on Discrete Mathematics and its Applications.
24. Gunn, S. R., (1998). Support vector machines for classification and regression. *ISIS Technical Report, 14*(1), 5–16.
25. He, S., Wu, Q. H., Wen, J. Y., Saunders, J. R., & Paton, R. C., (2004). A particle swarm optimizer with passive congregation. *Biosystems, 78*(1–3), 135–147.
26. Hirschberg, D. S., & Wong, C. K., (1976). A polynomial-time algorithm for the knapsack problem with two variables. *Journal of the ACM (JACM), 23*(1), 147–154.
27. Hu, X., Eberhart, R. C., & Shi, Y., (2003). Particle swarm with extended memory for multiobjective optimization. In: *Swarm Intelligence*

Symposium, 2003, SIS'03; Proceedings of the 2003 IEEE (pp. 193–197). IEEE.

28. Kågström, B., & Van, L. C., (1998). Algorithm 784: GEMM-based level 3 BLAS: Portability and optimization issues. *ACM Transactions on Mathematical Software (TOMS), 24*(3), 303–316.
29. Kannan, R., (1980). A polynomial algorithm for the two-variable integer programming problem. *Journal of the ACM (JACM), 27*(1), 118–122.
30. Karp, R. M., (1986). Combinatorics, complexity, and randomness. *Commun. ACM, 29*(2), 97–109.
31. Kennedy, J., & Mendes, R., (2002). Population structure and particle swarm performance. In: *Evolutionary Computation, 2002. CEC'02; Proceedings of the 2002 Congress* (Vol. 2, pp. 1671–1676). IEEE.
32. Kennedy, J., & Mendes, R., (2006). Neighborhood topologies in fully informed and best-of-neighborhood particle swarms. *IEEE Transactions on Systems, Man, and Cybernetics, Part C (Applications and Reviews), 36*(4), 515–519.
33. Kennedy, J., Eberhart, R. C., & Shi, Y., (2001). *Swarm Intelligence*. Morgan Kaufmann Publishers. Inc., San Francisco, CA.
34. Kröse, B., Krose, B., Van, D. S. P., & Smagt, P., (1993). *An Introduction to Neural Networks*. (Vol.1, pp.5-10)
35. Kukuk, M., (1997). *Kompakte Antwortdatenbasen Fiur die Liosung Von Geometrischen Anfrageproblemen Durch Abtastung*. Doctoral dissertation, Diplomarbeit, Informatik VII, universitiat Dortmund.
36. Li, J., (2004). *PeerStreaming: A Practical Receiver-Driven Peer-to-Peer Media Streaming System*. Microsoft Research MSR-TR-2004-101, Tech. Rep.
37. Liu, H., Luo, P., & Zeng, Z., (2007). A structured hierarchical P2P model based on a rigorous binary tree code algorithm. *Future Generation Computer Systems, 23*(2), 201–208.
38. Nisan, N., & Wigderson, A., (1995). On the complexity of bilinear forms: Dedicated to the memory of Jacques Morgenstern. In: *Proceedings of the Twenty-Seventh Annual ACM Symposium on Theory of Computing* (pp. 723–732). ACM.
39. Panda, P. R., Nakamura, H., & Dutt, N. D., (1997). Tiling and data alignment. *Solving Irregularly Structured Problems in Parallel Lecture Notes in Computer Science*. (Vol.1, pp.10-15)

40. Pereira, M., (2009). Peer-to-peer computing. In: *Encyclopedia of Information Science and Technology,* (2nd edn., pp. 3047–3052). IGI Global.
41. Pham, D., & Karaboga, D., (2012). *Intelligent Optimization Techniques: Genetic Algorithms, Tabu Search, Simulated Annealing and Neural Networks*. Springer Science & Business Media.
42. Rabin, M. O., (1977). Complexity of computations. *Communications of the ACM, 20*(9), 625–633.
43. Ratakonda, K., & Turaga, D. S., (2008). Quality models for multimedia delivery in a services-oriented architecture. *Managing Web Service Quality: Measuring Outcomes and Effectiveness: Measuring Outcomes and Effectiveness*.
44. Restrepo, J. H., Sánchez, J. J., & Hoyos, M. M. A. R. I. O., (2004). Solución al problema de entrega de pedidos utilizando recocido simulado. *Scientia et Technica, 10*(24).
45. Roucairol, C., (1996). Parallel processing for difficult combinatorial optimization problems. *European Journal of Operational Research, 92*(3), 573–590.
46. Sevkli, M., & Aydin, M. E., (2006). A variable neighborhood search algorithm for job shop scheduling problems. In: *European Conference on Evolutionary Computation in Combinatorial Optimization* (pp. 261–271). Springer, Berlin, Heidelberg.
47. Sevkli, M., & Guner, A. R., (2006). A continuous particle swarm optimization algorithm for uncapacitated facility location problem. In: *International Workshop on Ant Colony Optimization and Swarm Intelligence* (pp. 316–323). Springer, Berlin, Heidelberg.
48. Shi, Y., (2001). Particle swarm optimization: Developments, applications and resources. In: *Evolutionary Computation, 2001; Proceedings of the 2001 Congress* (Vol. 1, pp. 81–86). IEEE.
49. Skorin-Kapov, J., (1990). Tabu search applied to the quadratic assignment problem. *ORSA Journal on Computing, 2*(1), 33–45.
50. Stockmeyer, L., & Meyer, A. R., (2002). Cosmological lower bound on the circuit complexity of a small problem in logic. *Journal of the ACM (JACM), 49*(6), 753–784.
51. Stockmeyer, L., (1987). Classifying the computational complexity of problems. *The Journal of Symbolic Logic, 52*(1), 1–43.

52. Taillard, E. D., (1995). Comparison of iterative searches for the quadratic assignment problem. *Location Science, 3*(2), 87–105.
53. Taillard, É., (1991). Robust taboo search for the quadratic assignment problem. *Parallel Computing, 17*(4, 5), 443–455.
54. Thallner, B., & Moser, H., (2005). Topology control for fault-tolerant communication in highly dynamic wireless networks. In: *Intelligent Solutions in Embedded Systems, 2005; Third International Workshop* (pp. 89–100). IEEE.
55. Tran, D. A., Hua, K. A., & Do, T. T., (2004). A peer-to-peer architecture for media streaming. *IEEE Journal on Selected Areas in Communications, 22*(1), 121–133.
56. Vapnik, V., (2013). *The Nature of Statistical Learning Theory*. Springer Science & Business Media.
57. Vitter, J. S., & Simons, R. A., (1986). New classes for parallel complexity: A study of unification and other complete problems for P. *IEEE Transactions on Computers, 35*(5), 403–418.
58. Wierzbicki, A., Strzelecki, R., Swierezewski, D., & Znojek, M., (2002). Rhubarb: A tool for developing scalable and secure peer-to-peer applications. In: *Peer-to-Peer Computing, 2002 (P2P 2002); Proceedings Second International Conference* (pp. 144–151). IEEE.
59. Wilson, G. V., & Pawley, G. S., (1988). On the stability of the travelling salesman problem algorithm of Hopfield and tank. *Biological Cybernetics, 58*(1), 63–70.
60. Wu, X., Sharif, B. S., & Hinton, O. R., (2005). An improved resource allocation scheme for plane cover multiple access using genetic algorithm. *IEEE Transactions on Evolutionary Computation, 9*(1), 74–81.
61. Xiang, Z., Zhang, Q., Zhu, W., & Zhang, Z., (2003). Replication strategies for peer-to-peer based multimedia distribution service. In: *Multimedia and Expo, 2003, ICME '03; Proceedings 2003 International Conference* (Vol. 2, pp. II–153). IEEE.
62. Xiang, Z., Zhang, Q., Zhu, W., Zhang, Z., & Zhang, Y. Q., (2004). Peer-to-peer based multimedia distribution service. *IEEE Transactions on Multimedia, 6*(2), 343–355.
63. Xie, Z. P., Zheng, G. S., & He, G. M., (2006). Efficient loss recovery in application overlay stored media streaming. In: *Visual Communications*

and Image Processing 2005 (Vol. 5960, p. 596008). International Society for Optics and Photonics.

64. Yigit, V., Aydin, M. E., & Turkbey, O., (2004). Evolutionary simulated annealing algorithms for uncapacitated facility location problems. In: *Adaptive Computing in Design and Manufacture VI* (pp. 185–194). Springer, London.
65. Yigit, V., Aydin, M. E., & Turkbey, O., (2006). Solving large-scale uncapacitated facility location problems with evolutionary simulated annealing. *International Journal of Production Research, 44*(22), 4773–4791.

CHAPTER 3

FUNDAMENTALS OF SEARCH ALGORITHMS

CONTENTS

3.1. INTRODUCTION

Relational databases have evolved as a powerful technical tool for data transfer and manipulation over the last few years. Rapid improvements in data technology and science have had a substantial influence on how data is represented in current years (Abiteboul et al., 1995, 1997, 1999). In database technology, a new issue has emerged that restricts the effective representations of the data using classical tables. Data are shown as graphs and trees in several database systems. Certain applications, on either side, need a specific database system to manage time and uncertainty. To overcome the issues faced in database systems, significant research is being done (Adalı and Pigaty, 2003; Shaw et al., 2016).

At the moment, the researchers are looking at the possibility of developing models for "next-generation database systems," or databases that may reflect new data kinds and provide unique manipulation capabilities while still enabling regular search operations. Next-generation databases effectively handle structured documents, Web, network directories, and XML by modeling data in the format of trees and graphs (Figure 3.1) (Altınel and Franklin, 2000; Almohamad and Duffuaa, 1993).

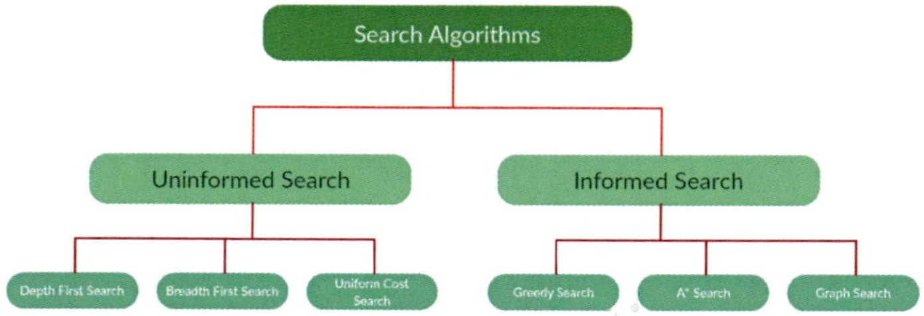

Figure 3.1. Search algorithms categorization.

Source: https://www.geeksforgeeks.org/search-algorithms-in-ai/.

In addition, modern database system utilizes tree or graph models for the representation of data in a variety of applications, including picture databases, molecular databases, and commercial databases. Due to the enormous importance of tree and graph database systems, several methods for tree and graph querying are now being developed (Altman, 1968; Amer-Yahia et al., 2001, 2002). Certain applications, including databases for commercial package delivery, meteorological databases, and financial databases, require temporal uncertainty coupled with database items in addition to querying and analyzing trees and graphs (Atkinson et al., 1990; Andries and Engels, 1994).

In the discipline of computer science, effective searching and sorting are considered essential and widely encountered difficulties. For example, the goal of a search algorithm for the gathering of things is to discover and differentiate a specific object from the rest. The search algorithm, on either side, may do a recognition analysis of a certain object that does not exist in the system (Baeza-Yates, 1989; Baeza-Yates et al., 1994). Database objects frequently contain key values that serve as the foundation for a search. Furthermore, certain data values contain information that will be obtained when an item is discovered (Baeza-Yates and Gonnet, 1996; Baeza-Yates and Ribeiro-Neto, 1999). A telephonic book, for example, provides a list of contacts with various contact information. Following the inclusion of some search inputs, the search algorithm is utilized to find specific contact details. It is common knowledge that certain data, such as a name or a number, is connected with such key values. Consider a search scenario in which you're looking for a single key-value pair (such as name). A list or an array is frequently used to hold the collection of things. The ith element (i.e., A[i]) normally correlates to the key value for the ith item present in a collection of *n* (number of) objects in a specific array A (i.e., A [1, …, n]) (Barrow and Burstall, 1976; Barbosa et al., 2001; Boag et al., 2002).

The objects are frequently sorted utilizing key values (for example, in a phone book), but this does not always necessary. Based upon the data's sorting condition, several search methods can be necessary to find the information (such as either sorted or not sorted). For a specific search algorithm, the inputs are the number of objects (i.e., n), an array of items (i.e., A), and the key-value pairs to be used in locating the objects (i.e., x). This explains the many kinds of search algorithms that are available (Boncz et al., 1998; Bomze et al., 1999).

3.2. UNORDERED LINEAR SEARCH

Consider the assumption that every given array is not fundamentally ordered. This could correlate to unsorted collection evaluations that do not have alphabetical sorting in the first place. As an example, what would be the best way for a student to acquire her or his exam results? To find his or her exam, she or he would look through the entire collection of examinations in a chronological manner (Boole, 1916; Bowersox et al., 2002). This search is related to algorithms that use unordered linear search results (Figure 3.2).

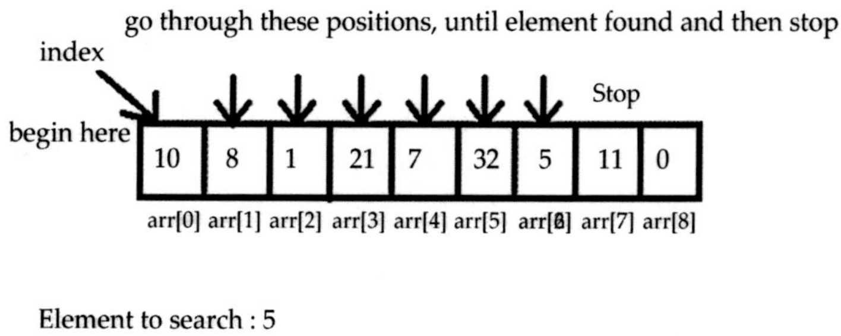

Figure 3.2. Linear search explained in simple terms.

Source: https://medium.com/karuna-sehgal/an-simplified-explanation-of-linear-search-5056942ba965.

The following is a typical case of an Unordered Linear Search:

- **Input:** An objects array; n shows the number of objects; the x-key value is being determined.
- **Output:** return point "*i,* " if not, return note "x not found"
- Perform the x comparison with every array (A) element A from the beginning;
- If x is the ith element in array A, returns point "i" once the search is completed;
- If not, keep looking for the next entry till the array is finished;
- Return note "x not found" if any suitable element is not found in the array (A).

To confirm the presence of a specific object, you must search the whole collection. Let's have a look at the following array:

i	I	II	III	IV	V	VI	VII	VIII
A	35	17	26	34	8	23	49	9

For example, if we require to look for the value x = 34 in this array (A). We require a comparison of x with (35, 17, 26, 34), with every element being compared just once each. Position 4 contains the required number (34) which is present. Thus, after four comparisons have been performed, we return 4.

If the search for the value x = 19 in this specific array is necessary. To complete this task, we must compare x with every one of the following elements: (35, 17, 26, 34, 8, 23, 49, 9), every element once. After going through all of the objects in this array, we are unable to locate number 19. Hence, we return "18 not found." In this particular situation, we have carried out a total of eight comparisons. Most of the time, it is necessary to do a determination (search) of x in an array of unordered objects that has n entries. It may be necessary to search through the entire array to find the appropriate response in some cases. It entails the implementation of a set of n contrasts. It is possible to create the following equation to express the number (n) of executed comparisons: T (n) = n (Brin, 1995; Bozkaya and Ozsoyoglu, 1999).

3.3. ORDERED LINEAR SEARCH

Assume that the array you're given is sorted. In these circumstances, a search across the entire list to identify a specific object or enquire about its existence in the collection of objects is not required. For example, if a collection of test results is sorted by name, it is not necessary to look beyond the "J"s to see if the exam score for "Jacob" is included in the collection or not included. The ordered linear search algorithms are the outcome of a simple modification of the algorithm (Brusoni et al., 1995; Console et al., 1995; Buneman et al., 1998). The following is an example of an Ordered Linear Search model:

- **Input:** B-objects array; n-number of objects; determining the x-key value.
- **Output:** return point "*i*," if not, return note "x not found"
- From the beginning of the array B, perform the comparison of x with the element B[i] in A to confirm their equality;
- If x = B[i], stop the search and return location i;
- If not, execute the comparison of x with that element once again too if the value of x is larger than B[i];
- If x > B[i], continue looking for the next object in array B;
- If not (such as x is less than B[i]), stop the search and return a message "x not found."

Take a look at the sorted version of the earlier utilized array:

i	I	II	III	IV	V	VI	VII	VIII
B	8	9	17	23	26	34	35	49

When looking for x = 34 in the earlier-mentioned array, the comparison of x with each element (8, 9, 17, 23, 26) is done twice (such as once for "=" and again for ">"). The comparison of x with (34) is then performed just once for "=." As a result, we'll locate 34 in Position 6 and it'll return to Position 6. $2 \times 5 + 1 = 11$ is the total number of comparisons.

When searching this array for x = 19, the comparison of x with each member (8, 9, 17, 23, 26) is done twice (such as once for "=" and again for ">"). And in the last comparison (such as if x > 25), we receive a "NO" result, implying that all of the objects in this array after 26 are effectively bigger than x. As a result, the message "x not found" is returned. $2 \times 4 = 8$ is the total number of comparisons.

In general, determining (searching) for x in an array of ordered objects with n objects is necessary. When the value of x is larger than all of the array's values, it's often necessary to search the entire array for the required response. It entails executing $n \times 2$ comparisons (such as n for "=" and another n for ">"). T (n) = 2n is an equation that represents the number (n) of executed comparisons.

3.4. CHUNK SEARCH

There is no requirement to look through the entire collection chronologically in the scenario of an ordered list. Suppose identifying a name in a phone book or a specific exam in a sorted collection: one may immediately select 40 or more pages from the phone book or 20 or more examinations from the collection to easily identify the 40-page (or 20 exams) chunk (pile) wherein the needed information is included. An organized linear search technique must be used to carefully sift through this pile (chunk). Suppose that c is the chunk size for 40 pages or 20 examinations. Furthermore, we might suppose that we have access to a much more generic method for ordered linear search. Such assumptions, when combined with the previously discussed notions, may lead to the development of the chunk search algorithm (Burns and Riseman, 1992). The following is an example of a common Chunk search algorithm:

- **Input:** An A-ordered array of objects, c-chunk size, n-the number of elements, determining the x-key value.
- **Output:** if found, return position "*i*," if not, return the message "x not found." Disintegrate array A into c-sized chunks.
- Compare x with the last components of every chunk, except the last chunk;
- Check to see if the value of x is larger than that element;

- If affirmative, go to the next chunks;
- If no, it signifies that x is in that chunk;
- Within the chunk, run the ordered linear search algorithm.

Take a look at the following array:

i	I	II	III	IV	V	VI	VII	VIII
B	8	9	17	23	26	34	35	49

Let's say the chunk size is 2 and we're looking for x = 34. We begin by dividing the array into 4 chunks of size 2, then doing a one-time comparison of x with the final piece of every element (9, 23, 34). We compare the values of x and that element to check if the value of x is bigger. Whenever we ran the comparison for 34, we got the result "NO," implying that x must be in the 3rd chunk.

Ultimately, in the 3rd chunk, an ordered liner search is performed, followed by the location of 33 at position 6. In this instance, we run three comparisons to select the appropriate chunk, then three more comparisons within the chunk (such as 2 for 26 and 1 for 34). In total, 6 comparisons are made. In general, searching for x in an array of ordered objects with chunk sizes of c and n items is required (Chase, 1987). In the worst-case situation, n/c – 1 comparison are performed to select the correct chunk, whereas 2c comparisons are performed to do the linear search. T(n) = n/c +2c – 1 is an equation that may be used to describe the number (n) of executed comparisons. We usually ignore the constant number since it has no effect as n increases. Ultimately, T(n) = n/c +2c is obtained (Cai et al., 1992; Caouette et al., 1998).

3.5. BINARY SEARCH

Consider the following notion for a search algorithm using the phone book as an instance. Assume we pick a page from the center of a phonebook at random. We've succeeded if the name we're looking for is listed on this page (Figure 3.3).

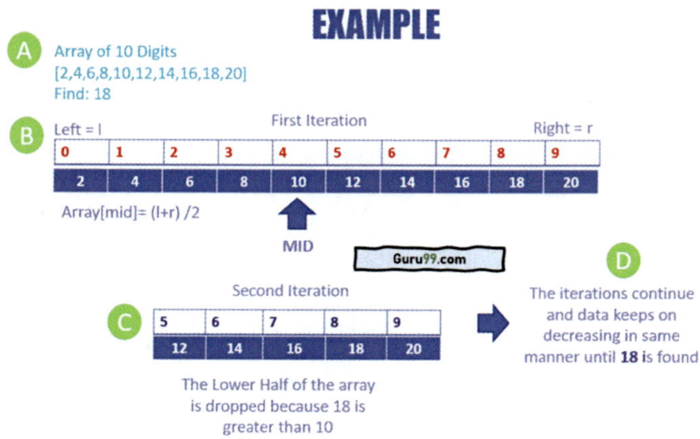

Figure 3.3. Binary search algorithm with an example.

Source: https://www.guru99.com/binary-search.html.

The operations are repeated on the 1st half of the phonebook if the specific name being determined appears alphabetically before this page; alternatively, the processes are repeated on the 2nd half of the phonebook. It should be noted that every iteration entails splitting the remaining piece of the phonebook to be searched into 2 halves; this approach is referred to as binary search (Chiueh, 1994; Christmas et al., 1995; Ciaccia et al., 1997). Although this technique can not appear to be the best for scanning a phonebook (or an ordered list), it is likely the fastest. This is true for several computer algorithms; the most natural (appropriate) algorithm isn't always the best (Cole and Hariharan, 1997). The following is an example of a Binary Search algorithm model:

- **Input:** determining the "x" key-value, n-the number of elements, an A-ordered array of objects.
- **Output:** if found, return position "*i*," if not, return the message "x not found."
- Split the array into 2 equal halves;
- Check equivalence of x with the 1st half's final element by comparing it to that element;
- If so, stop looking and return to your original location;
- If no, compare the value of x to the final element of the 1st half to check if the value of x is larger than that element;

- If yes, x must be in the 2nd part of the equation. The 2nd half of the array should now be handled as a new array to run the Binary Search on it;
- If no, then x should be placed in the 1st half. The 1st half of the array should now be handled as a new array, and Binary Search should be run on it;
- If x isn't discovered, display a notice that says "x not found."

Consider the following depicted array once more:

i	I	II	III	IV	V	VI	VII	VIII
B	8	9	17	23	26	34	35	49

If we require to find x = 26. We shall compare x to every element (23, 34) two times (such as one for "=" and another for ">"), followed by a single comparison to element (26) for "=." Ultimately, at the fifth place, we discover x. There are a total number of comparisons: $2 \times 2 + 1 = 5$. T (n) = 2log2n in the worst-case situation (Cook and Holder, 1993; Cole et al., 1999).

3.6. SEARCHING IN GRAPHS

Due to their broad, powerful, and flexible form, graphs are commonly employed to describe data in a variety of applications. A graph is made up of a collection of vertices and edges that connect the pairs of vertices. Generally, edges are utilized to depict the relationships between various data variables, and vertices are utilized to depict the data itself (such as anything which needs a description). Based upon the level of abstraction used to describe the information (data), a graph illustrates the data either semantically or syntactically (Figures 3.4–3.6).

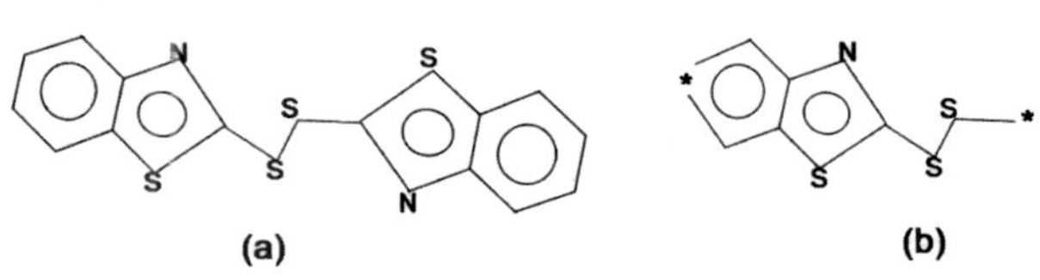

Figure 3.4. (a) The chemical formula of a compound. (b) A query consisting of wildcards. Graphs are naturally utilized to explain their structures.

Source: https://www.semanticscholar.org/paper/Searching-Algorithms-and-Data-Structures-for-%2C-and-Giugno/4378f3b5d0495f4164c4ac74f01ecef414951974.

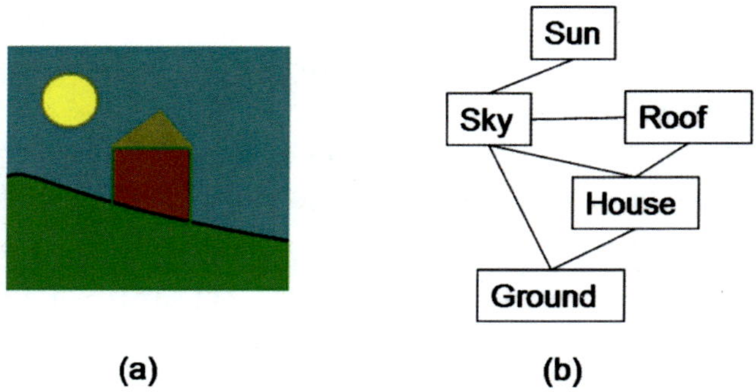

Figure 3.5. (a) Representation of an image; (b) illustration of a region adjacent graph – RAG of the image.

Source: https://www.semanticscholar.org/paper/Searching-Algorithms-and-Data-Structures-for-%2C-and-Giugno/4378f3b5d0495f4164c4ac74f01ecef414951974.

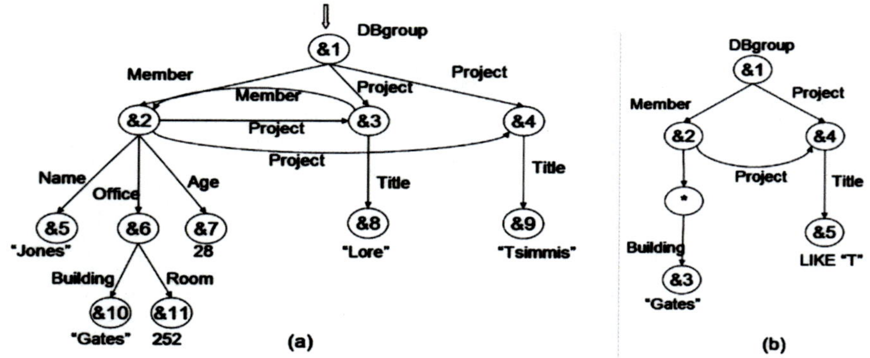

Figure 3.6. Illustration of (a) a structured database tree; and (b) a query containing wildcards.

Source: https://www.semanticscholar.org/paper/Searching-Algorithms-and-Data-Structures-for-%2C-and-Giugno/4378f3b5d0495f4164c4ac74f01ecef414951974.

Let's look at a biological database system (for example, proteins). Commonly, proteins are described utilizing labeled graphs, in which the edges reflect the links between distinct atoms and the vertices show specific atoms. Proteins are often classified depending upon their shared structural characteristics. Estimating the functioning of a novel protein fragment is one of the uses of these classifications (such as synthesized or discovered). The functioning of the novel fragments may be deduced by looking for structural similarities between the novel fragments and known proteins. Furthermore, wildcards can be present in searches that have matching properties with the data's vertices or pathways (Corneil and Gotlieb, 1970; Day et al., 1995).

Graphs are used as the fundamental data structures in visual languages. Such languages are used in computer science and software engineering to design projects and tools for integrated environments, in CIM systems to display process modeling, and in visual database systems to describe query language semantics (Bancilhon et al., 1992; Dehaspe et al., 1998; Dekhtyar et al., 2001).

Computer vision graphs are utilized to depict a variety of pictures at various degrees of abstraction (DeWitt et al., 1994; Deutsch et al., 1999). The graph vertices relate to edges, and pixels to spatial relationships between pixels, in a depiction with the low-level description (Djoko et al., 1997; Dinitz et al., 1999). The picture depicted by a graph (that is, RAG) is carried out at higher description levels in a way that image breakdown in areas occur. The areas in this case reflect the graph vertices, while the edges reflect the geographical relationships between them (Dutta, 1989; Dubiner et al., 1994).

Graphs are commonly used to model data in semi-structured database systems, network directories, and the Web (Dyreson and Snodgrass, 1998). These database systems generally comprise directed labeled graphs with complicated items at the vertices and links between the objects represented by the edges. The huge size of these database systems necessitates the use of wildcards in query specification. Such a phenomenon enables the retrieval of sub-graphs using just a portion of the graph's data (Eshera and Fu, 1984; Engels et al., 1992; Cesarini et al., 2001).

Besides storing data in graphs, most of the applications listed above need the usage of tools for comparing graphs, identifying distinct sections of graphs, obtaining graphical data, and categorizing data. Recent search engines provide quick responses for keyword-based searches in the case of non-structured data (for example, strings). Caching, inverted (smart) index structures and the utilization of parallel and distributed computing are only a few of the variables that contribute to the high speed (Ferro et al., 1999, 2001; Foggia et al., 2001). The query graphs are compared against the core data graphs in key graph searching, identical to how words are matched in keyword searching. There have been considerable efforts to generalize keyword searches for key graph searching (Frakes and Baeza-Yates, 1992; Fortin, 1996). Furthermore, such generalizations are not entirely natural; for example, keyword searching has exponential complexity as a function of database size. Key graph searching, on either side, has exponential complexity, making it a different class of issues. The next sections cover the many sorts of issues connected with key graph searching (Gadia et al., 1992; Fernandez et al., 1998; Fernández and Valiente, 2001).

3.6.1. Exact Match or Isomorphisms

We can tell if two graphs are the same if we have a data graph G_b and a query graph G_a. The isomorphic nature of the graphs G_b and G_a are determined, as well as the mapping of G_b vertices and G_a vertices while keeping the conforming edges in G_b and G_a. This problem is known to be in NP (nondeterministic polynomial time), and whether it is in NP-complete or P (polynomial time) is uncertain (Figure 3.7) (Garey and Johnson, 1979; Gold and Rangarajan, 1996; Goldman and Widom, 1997).

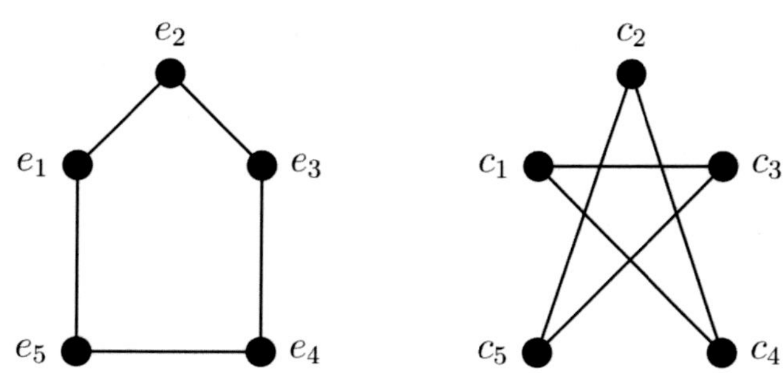

Figure 3.7. Instances of isomorphic graphs.

Source: https://math.stackexchange.com/questions/3141500/are-these-two-graphs-isomorphic-why-why-not.

3.6.2. Subgraph Exact Matching or Subgraph Isomorphism

We may readily suppose that G_a is a subgraph isomorphic to G_b given a data graph G_b and a query graph G_a, assuming that G_a is also isomorphic to the subgraph of G_b. It's worth noting that G_a has the potential to be subgraph isomorphic for a variety of G_b subgraphs. That problem is thought to be NP-complete. Furthermore, instead of identifying a single instance of Gagraph in G_b, it is far more costly tc locate all the subgraphs that demonstrate similarity with the query graph G_a (for example, demonstrating the most occurrences of Gagraph in G_b) (Gonnet and Tompa, 1987; Grossi, 1991, 1993).

3.6.3. Matching of Subgraph in a Graphs' Database

We wish to find all the G_a occurrences in every graph of D given a data graph D and a query graph G_a. While graph-to-graph matching algorithms may be used, particular strategies have shown to be effective in reducing the time complexity and search space in database systems. This issue is also NP-complete (Hirata and Kato, 1992; Güting, 1994; Gupta and Nishimura, 1998).

The construction of all possible maps in-between vertices of the two graphs, accompanied by a verification of the map's matching attributes, is a basic listing method for discovering the existence of the query graph G_a in the data graph G_b. The algorithms with these networks had exponential complexity (Hirschberg and Wong, 1976; Kannan, 1980; Goodman and O'Rourke, 1997).

There have been several attempts to reduce the combinatorial costs of graph searching. The following three suggestions for research have been pursued in this domain:

- The 1st effort is to investigate matching algorithms for particular graph structures, such as associated graphs, planar graphs, and bounded valence graphs (Umeyama, 1988; Cour et al., 2007);
- The 2nd effort is to investigate mechanisms for decreasing the number (quantity) of generated maps (Wilson and Hancock, 1997; Luo and Hancock, 2001); and
- Ultimately, the 3rd research effort is to provide approximate polynomial-complexity methods; nevertheless, these algorithms don't guarantee a proper solution (Leordeanu and Hebert, 2005; Leordeanu et al., 2009).

For key graph searching, a variety of algorithms are developed to cope with situations in which accurate matches are hard to locate (Milo and Suciu, 1999; Conte et al., 2004). These kinds of algorithms are extremely effective in situations that include noisy graphs. Such algorithms generally make use of a cost function to forecast the similarity of two graphs and to accomplish the conversion of two graphs into each other, respectively (McHugh et al.,

1997; Al-Khalifa et al., 2002; Chung et al., 2002). For example, *semantic transformations* may be utilized to construct a cost function that is primarily based upon the particular application domains and to allow the vertices to match with discordant values when the application domains are different. Additionally, syntactic changes (such as branch deletion and insertion) are required for matching structurally dissimilar regions of the graphs, and these changes are reliant on *semantic transformations* as well. In the case of noisy data graphs, approximation techniques can also be used as an alternative (Ciaccia et al., 1997; Cooper et al., 2001; Kaushik et al., 2002).

In the scenario of query graphs that are present in the database of graphs, the majority of modern approaches are designed for particular purposes in mind (Gyssens et al., 1989; Salminen and Tompa, 1992). Different researchers have suggested numerous querying techniques for semi-structured database systems. Furthermore, a large number of commercial products and research studies have been completed that use subgraph searching in various biochemicals database systems (Macleod, 1991; Kilpeläinen and Mannila, 1993). These two separate instances have underlying data models that are distinct from one another (such as initially the database is seen as a large graph in case of commercial products whereas the database is seen as a collection of graphs in case of academic projects). The strategies outlined above, on the other hand, demonstrate the following common approaches:

- Regular path expressions;
- Regular indexing methods.

These approaches are utilized during query time to discover the database's substructures as well as avoid needless database traversals (Tague et al., 1991; Navarro and Baeza-Yates, 1995).

Just a few application-independent strategies exist for querying graph database systems, compared to application-specific alternatives. In most database systems, query graphs of the identical size are used; although, certain approaches allow the same-size limitation to be restricted (Dublish, 1990; Burkowski, 1992; Clark et al., 1995). The algorithms are based on the development of a similarity index between the database's subgraphs and graphs, accompanied by their arrangement of appropriate data structures. Bunke (2000) suggested a technique for indexing labeled database graphs in exponential time and computing the isomorphism of a subgraph in polynomial time. Matching and indexing are both dependent upon all possible permutations of the graphs' neighboring matrices. If only a set of plausible permutations is kept in mind, the aforementioned method may perform better (Verma and Reyner, 1989; Nishimura et al., 2000). Cook and Holder (1993) proposed an alternative way for looking for a subgraph in a database that was not reliant on any indexing mechanism. After applying traditional graph matching techniques to a single-graph database system, they discovered comparable recurring subgraphs (Luccio and Pagli, 1991; Fukagawa and Akutsu, 2004).

3.7. GRAPH GREP

Here we will explain how to use a graph database system to execute precise subgraph queries utilizing an application-independent technique. GraphGrep is a tool that searches a graph database for all potential instances of a specific graph (Güting, 1994; Ganapathysaravanabavan and Warnow, 2001). The indexing techniques that categorize minor substructures of the graphs existing in a database are frequently used in search algorithms employing application-independent methods. The graph vertices of GraphGrep have a label (label-vertex) and an identification number (such as id-vertex). We may suppose that graph labeling occurs solely at the vertices in this case. An empath of length n is a series of n + 1 id-vertices with a binary relationship between any two consecutive vertices. A label route of length n, on the other hand, depicts a succession of n + 1 label vertices (Figure 3.8) (Gupta and Nishimura, 1995; Kao et al., 1999).

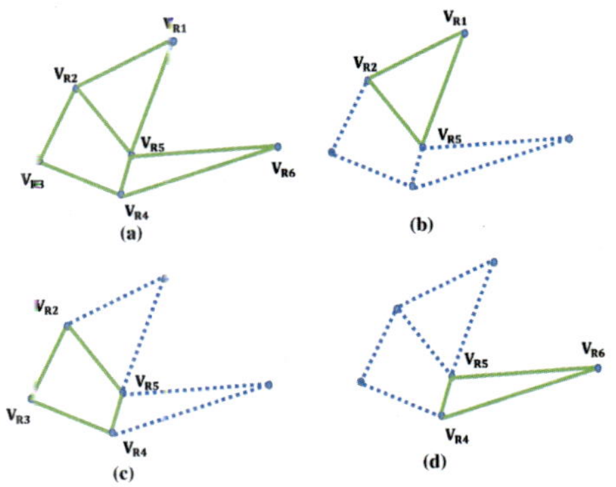

Figure 3.8. (a) A graph (GRep) with 6 vertices and 8 edges; (b, c, and d) possible cliques of GRep is $D_1 = \{V_{R1}, V_{R2}, V_{R5}\}$; $D_2 = \{V_{R2}, V_{R3}, V_{R4}, V_{R5}\}$, and $D_3 = \{V_{R6}, V_{R5}, V_{R4}\}$.

Source: https://www.researchgate.net/figure/a-A-graph-GRep-with-6-vertices-and-8-edges-b-c-and-d-possible-cliques-of-GRep-is_fig9_330763767.

Database fingerprints are the names given to the indexes. They are often built during the database's preparation step and serve as an abstract representation of the graphs' structural properties. The fingerprints are implemented utilizing a hash table, whereby each row displays the number of id-paths that are associated with the label path hashed in that row (Consens and Mendelzon, 1990; Kato et al., 1992; Hlaoui and Wang, 2002). Label pathways are normally 0 in length and have a constant value, in other words, lp. The pre-processing feature of the graphs can be executed in polynomial time with the right lp value. The id-paths created during fingerprinting are normally maintained rather than discarded; nonetheless, they are kept in tables, with every table representing a different label path. The data provided in tables is used by algebra to discover a match for the query (Hoffmann and O'Donnell, 1982; Hong and Huang, 2001).

Graph Linear Description language is a graph query language that we present for query formulation. XPath for XML documents and Smart-Smiles for molecules are 2 query languages that Glide is derived from (Hopcroft and Wong, 1974; James et al., 2000). Smart is a query language for identifying components in Smiles databases, while SMILES is a language meant for coding molecules. Glide takes Smiles' cycle notation and optimizes it for usage in any graph application. Complicated path expressions are utilized in XPath to represent queries that contain both the matching conditions and the filter in the vertex notation. Graph expressions are used instead of path expressions in Glide (Jensen and Snodgrass, 1999; Kanza and Sagiv, 2001).

Scientists evaluated the algorithm's effectiveness on NCI datasets with up to 120,000 molecules and also random databases (Kilpeläinen, 1992; Kilpeläinen and Mannila, 1994). Furthermore, experts have compared GraphGrep to the most widely used tools (Frowns and Daylight) and have come up with really encouraging results. The website (www.cs.nyu.edu/shasha/papers/graphgrep) has a software version of GraphGrep and a demo.

3.8. SEARCHING IN TREES

Trees are specialized forms of graphs that are used in many applications to describe data. There are several tools for searching, storing, indexing, and retrieving sub-trees inside a set of trees. The phrase "key tree searching" refers to a set of rooted tree sub-graph and graph matching techniques (Hlaoui and Wang, 2002, 2004; Drira and Rodriguez, 2009).

Take, for example, a database of old coins. The traditional technique of exchanging information about antique coins or other valuable ancient artifacts between archaeological institutes and museums is through photo collections. Generally, when a new coin is discovered, an expert applies his past knowledge and reviews all pertinent facts previously known about the object to determine the coin's origin and categorization (Pelegri-Llopart and Graham, 1988; Aho et al., 1989). Accessibility to coin databases, such as the one mentioned above, is a highly valuable tool for the confirmation or denial of any archaeological hypotheses during working on this difficult endeavor. Nonetheless, the quantity and size of catalogs accessible for consultation are restricted (Aho and Ganapathi, 1985; Karray et al., 2007). Computers are now altering this traditional framework in a variety of ways. To begin with, quick and low-cost scanning technology has greatly increased the number of available images. Secondly, "intelligent" procedures give essential assistance. When the size of picture databases reaches a certain point, traditional image search approaches become virtually ineffective. Algorithms for effectively evaluating and comparing hundreds of thousands of images signal a breakthrough in this direction (Hong et al., 2000; Hong and Huang, 2004). A coin is a complex object with a specific structure and syntax from a semantic standpoint. Its structure and syntax are defined by an expert by defining its most important traits, which aid in determining its identification. Such characteristics may be grouped into a tree, and the distance between two trees may be used as a heuristic way for estimating the distance between related coins (Luks, 1982; Xu et al., 1999). Figure 3.9 depicts a partial features tree for a generic coin. The arrangement of a tree-like structure is mostly based upon a detailed examination performed by an information technologist in collaboration with an archaeologist specialist. The more selective traits are associated with the higher-level nodes of a tree, according to a standard rule. This criterion, although, can be overridden if it will result in incompetence in dealing with the resulting tree structure (Figure 3.9) (Filotti and Mayer, 1980; Babai et al., 1982).

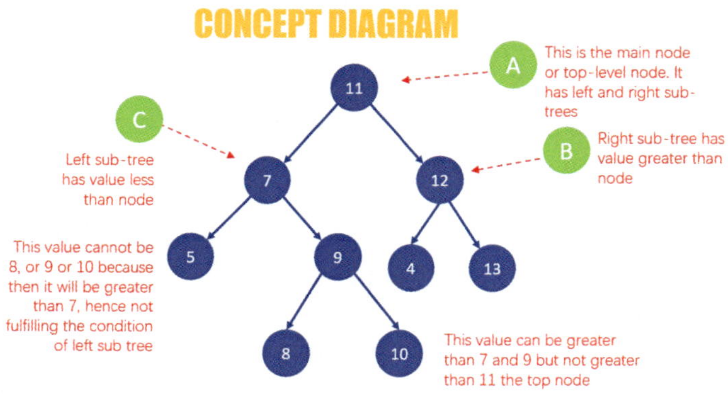

Figure 3.9. Attributes of binary search tree.

Source: https://www.guru99.com/binary-search-tree-data-structure.html.

XML is a standard language for exchanging and describing information on the internet that is becoming popular. This is due to the unique technique of reference between the components of XML, which results in the natural data structure of a graph being used. By omitting these references, an XML document is transformed into an ordered tree. The bulk of XML database systems have selected trees as their fundamental data model, and this is a good thing.

For the presentation of statements or the architecture of the document, trees are used in many applications of natural language processing (for instance, looking for matches in example-based translation systems and retrieving material from digital libraries). Authors from a variety of fields have established the hierarchical patterns of trees that describe the syntactic principles that govern the creation of English sentences. Furthermore, trees may be used to depict the geometrical aspects of document pages, which can be used to answer questions such as "find all of the pages that have the title next to an image" and "find all of the pages that have the title next to an image" (Jensen et al., 1998; Schlieder, 2002). In this situation, the hierarchical structures of trees correlate to such a portion in regions of pages where that portion in a page is an image or text between columns and white spaces, and the hierarchical structures of trees correlate to such a portion in regions of pages (Lueker and Booth, 1979).

Certain properties of the apps listed above are shared by all of them. We may show the database as either a single tree or as a collection of trees in our diagram. The order of siblings in a tree may also be crucial (as in the XML data format), or a tree may be unordered (as in certain archeological databases and hereditary trees), depending on the situation. In the same way that graphs require searching, finding an "approximate" or precise tree and sub-tree matching may be required. When working with approximate values, one approach of being accurate on the low end is to count the total number of routes in the query tree that do not appear in the data tree (Dolog et al., 2009). The matching with query trees that contain wildcards is included in the approximate tree matching (Navarro and Baeza-yates, 1995).

The complexity of key-tree searching issues varies based upon the structure of the tree and ranges from linear (P) to exponential (NP-complete) based upon the size of the tree. More precisely, the time required to solve an accurate sub-tree or tree matching issue is polynomial in both sorted and unordered trees, depending on the complexity of the challenge (Sikora, 2012). Estimated tree searching issues fall into the P class for ordered trees and the NP-complete class for unordered trees.

Extensive research efforts are undertaken to integrate different complex data structures with different approximation tree matching algorithms which work across generic metric spaces to reduce the processing time of the query (that is, algorithms that are not based upon specific characteristics of the distance function considered) (Figures 3.10–3.12).

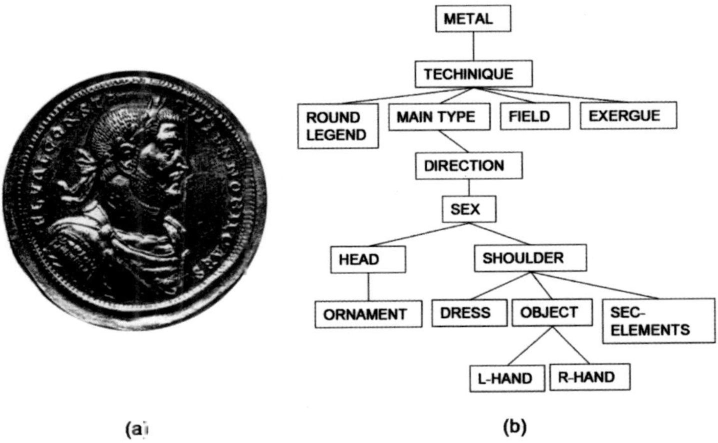

Figure 3.10. (a) A late Roman empire coin; (b) general tree diagram of a coin.

Source: https://www.semanticscholar.org/paper/Searching-Algorithms-and-Data-Structures-for-%2C-and-Giugno/4378f3b5d0495f4164c4ac74f01ecef414951974.

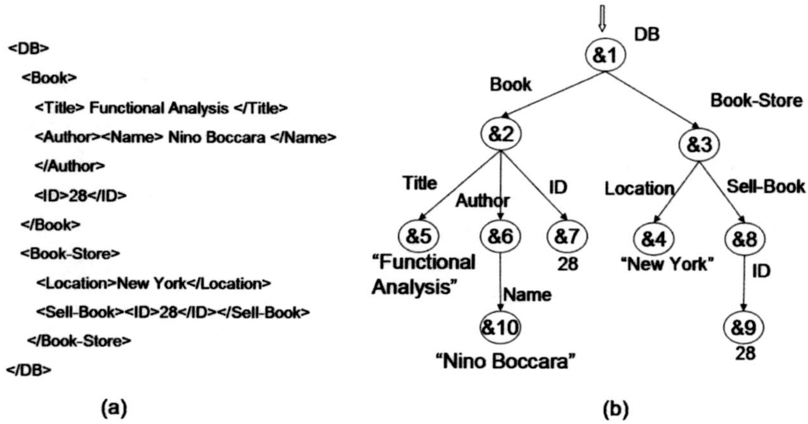

Figure 3.11. (a) An XML document; (b) an XML tree.

Source: https://www.semanticscholar.org/paper/Searching-Algorithms-and-Data-Structures-for-%2C-and-Giugno/4378f3b5d0495f4164c4ac74f01ecef414951974.

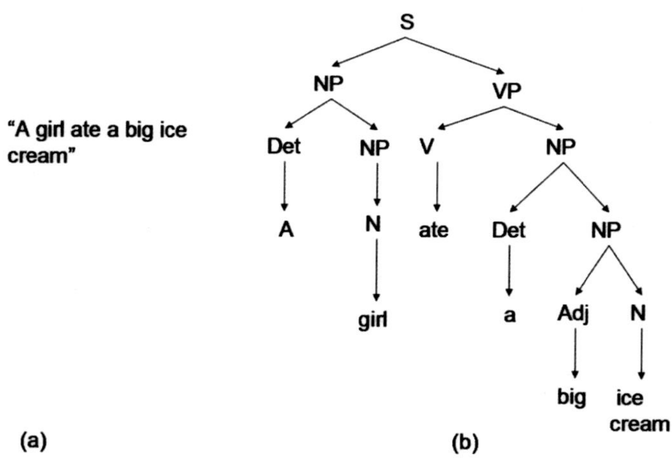

Figure 3.12. (a) An English sentence; (b) a tree elaborates the syntactic laws of the sentence.

Source: https://www.semanticscholar.org/paper/Searching-Algorithms-and-Data-Structures-for-%2C-and-Giugno/4378f3b5d0495f4164c4ac74f01ecef414951974.

Some tree-searching algorithms rely on the characteristics of the basic distance function between database items. Others merely consider the distance function to be a metric. The fixed query tree (FQ-tree) algorithm, the vantage point tree (VP-tree) method, its upgraded form (referred to as MVP-tree) algorithm, and the M-tree algorithm are all examples of this.

3.9. SEARCHING IN TEMPORAL PROBABILISTIC OBJECT DATA MODEL

After that, we will look at another component of next-generation databases: the specification of a temporal, probabilistic object database, which will be discussed in more detail later.

In a wide range of applications, object data models have been used to model them. Examples include financial risk applications, multimedia applications, logistics, and supply chain management systems and meteorological applications, amongst several others. Several of such applications are forced to express and handle both uncertainty and time as a matter of course (Figure 3.13).

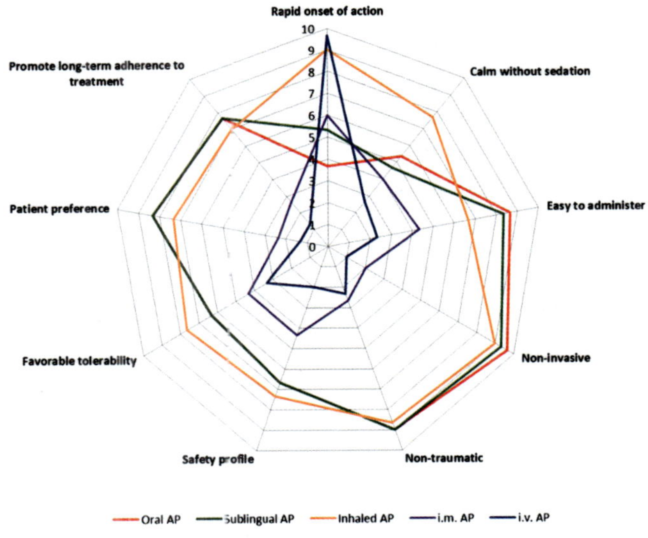

Figure 3.13. Temporal persistence modeling for object search.

Source: https://www.semanticscholar.org/paper/Temporal-persistence-modeling-for-object-search-Toris-Chernova/ef0e38a237c6159bd2547751bf-0014b9afcc2d9b/figure/2.

Firstly, we'll look at a logistics application for transportation. A commercial package delivery business (like DHL, FedEx, UPS, and others) possesses precise statistical data on how long it takes packages to travel from one zip code to another, and frequently even more particular data (such as how long it takes for a package from the address of one street to another street address). A company anticipating deliveries will like data of the form "the package would be transported between 1 p.m. and 5 p.m. with a probability between 0.8 and 0.9 and between 9 a.m. and 1 p.m. with a probability between 0.1 and 0.2" (here, probabilities are levels of belief about a future event, which can be derived from statistical data about identical previous events). The answer "It would be supplied sometime today between 9 a.m. and 5 p.m." is considerably less beneficial to the company's decision-making procedures than "It would be supplied sometime today between 9 a.m. and 5 p.m." For instance, it aids in the scheduling of workers, the preparation of receiving facilities (for toxic and other heavy products), the preparation of future production plans, and so on. Furthermore, object models have been normally utilized to store the various entities involved in an application, as various automobiles (trucks, airplanes, and so on.) have distinct characteristics and various packages (tube, letter, toxic material shipments for commercial clients, and so on.) have widely varying characteristics. The shipping firm itself has a high demand for this information. For instance, the corporation will need to query this information to build plans that best distribute and/or utilize current sources (staff, truck space, etc.), depending upon their estimates of future workload (Aho and Ganapathi, 1985; Karray et al., 2007).

Object models are used to express weather data in weather database systems (like the US Department of Defense's total Atmospheric and Ocean System, or TAOS). In weather models, uncertainty and time are ubiquitous and most decision-making algorithms depend upon this information to make judgments.

Banks and institutional lenders employ a variety of financial models to try to anticipate when clients would default on borrowing. Complicated mathematical models incorporating probabilities and time are examples of these models (the predictions specify the probability with which a given customer would default in a given time). In addition, models for predicting loan defaults and bankruptcies differ significantly based upon the market, the kind of credit instrument (mortgage, construction loan, customer credit card, HUD loan, commercial real estate loan, so on.), the factors that impact the loan, different aspects about the consumer, and so on. Object models are simply used to express these models, and uncertainty and time are used to parameterize different aspects of the model (Filotti and Mayer, 1980; Babai et al., 1982).

REFERENCES

1. Abiteboul, S., & Vianu, V., (1999). Regular path queries with constraints. *Journal of Computer and System Sciences, 58*(3), 428–452.
2. Abiteboul, S., Hull, R., & Vianu, V., (1995). *Foundations of Databases: The Logical Level.* Addison-Wesley Longman Publishing Co., Inc.
3. Abiteboul, S., Quass, D., McHugh, J., Widom, J., & Wiener, J. L., (1997). The lorel query language for semistructured data. *International Journal on Digital Libraries, 1*(1), 68–88.
4. Adalı, S., & Pigaty, L., (2003). The DARPA advanced logistics project. *Annals of Mathematics and Artificial Intelligence, 37*(4), 409–452.
5. Aho, A. V., & Ganapathi, M., (1985). Efficient tree pattern matching (extended abstract): An aid to code generation. In: *Proceedings of the 12th ACM SIGACT-SIGPLAN Symposium on Principles of Programming Languages* (pp. 334–340). ACM.
6. Aho, A. V., Ganapathi, M., & Tjiang, S. W., (1989). Code generation using tree matching and dynamic programming. *ACM Transactions on Programming Languages and Systems (TOPLAS), 11*(4), 491–516.
7. Al-Khalifa, S., Jagadish, H. V., Koudas, N., Patel, J. M., Srivastava, D., & Wu, Y., (2002). Structural joins: A primitive for efficient XML query pattern matching. In: *Data Engineering, 2002; Proceedings 18th International Conference* (pp. 141–152). IEEE.
8. Almohamad, H. A., & Duffuaa, S. O., (1993). A linear programming approach for the weighted graph matching problem. *IEEE Transactions on Pattern Analysis and Machine Intelligence, 15*(5), 522–525.
9. Altınel, M., & Franklin, M. J., (2000). Efficient filtering of XML documents for selective dissemination of information. In: *Proc. of the 26th Int'l Conference on Very Large Data Bases (VLDB).* Cairo, Egypt.
10. Altman, E. I., (1968). Financial ratios, discriminant analysis and the prediction of corporate bankruptcy. *The Journal of Finance, 23*(4), 589–609.
11. Amer-Yahia, S., Cho, S., & Srivastava, D., (2002). Tree pattern relaxation. In: *International Conference on Extending Database Technology* (pp. 496–513). Springer, Berlin, Heidelberg.
12. Amer-Yahia, S., Cho, S., Lakshmanan, L. V., & Srivastava, D., (2001). Minimization of tree pattern queries. In: *ACM SIGMOD Record* (Vol. 30, No. 2, pp. 497–508). ACM.

13. Andries, M., & Engels, G., (1994). Syntax and semantics of hybrid database languages. In: *Graph Transformations in Computer Science* (pp. 19–36). Springer, Berlin, Heidelberg.

14. Atkinson, M., DeWitt, D., Maier, D., Bancilhon, F., Dittrich, K., & Zdonik, S., (1990). The object-oriented database system manifesto. In: *Deductive and Object-Oriented Databases* (pp. 223–240).

15. Babai, L., Grigoryev, D. Y., & Mount, D. M., (1982). Isomorphism of graphs with bounded eigenvalue multiplicity. In: *Proceedings of the Fourteenth Annual ACM Symposium on Theory of Computing* (pp. 310–324). ACM.

16. Baeza-Yates, R. A., & Gonnet, G. H., (1996). Fast text searching for regular expressions or automaton searching on tries. *Journal of the ACM (JACM), 43*(6), 915–936.

17. Baeza-Yates, R. A., (1989). Algorithms for string searching. In: *ACM SIGIR Forum* (Vol. 23, No. 3, 4, pp. 34–58). ACM.

18. Baeza-Yates, R., & Ribeiro-Neto, B., (1999). *Modern Information Retrieval* (Vol. 463, pp. 1–20). New York: ACM Press.

19. Baeza-Yates, R., Cunto, W., Manber, U., & Wu, S., (1994). Proximity matching using fixed-queries trees. In: *Annual Symposium on Combinatorial Pattern Matching* (Vol. 1, pp. 198–212). Springer, Berlin, Heidelberg.

20. Bancilhon, F., Delobel, C., & Kanellakis, P., (1992). *Building an Object-Oriented Database System: The Story of 0 2*. Morgan Kaufmann Publishers Inc.

21. Barbosa, D., Barta, A., Mendelzon, A. O., Mihaila, G. A., Rizzolo, F., & Rodriguez-Gianolli, P., (2001). ToX-the Toronto XML engine. In: *Workshop on Information Integration on the Web* (pp. 66–73).

22. Barrow, H. G., & Burstall, R. M., (1976). Subgraph isomorphism, matching relational structures and maximal cliques. *Information Processing Letters, 4*(4), 83–84.

23. Boag, S., Chamberlin, D., Fernández, M. F., Florescu, D., Robie, J., Siméon, J., & Stefanescu, M., (2002). *XQuery 1.0: An XML Query Language*.(Vol.1, pp. 1-5)

24. Bomze, I. M., Budinich, M., Pardalos, P. M., & Pelillo, M., (1999). The maximum clique problem. In: *Handbook of Combinatorial Optimization* (pp. 1–74). Springer, Boston, MA.

25. Boncz, P., Wilshut, A. N., & Kersten, M. L., (1998). Flattening an object algebra to provide performance. In: *Data Engineering, 1998; Proceedings 14th International Conference* (pp. 568–577). IEEE.
26. Boole, G., (1916). *The Laws of Thought* (Vol. 2, pp. 1–20). Open Court Publishing Company.
27. Bowersox, D. J., Closs, D. J., & Cooper, M. B., (2002). *Supply Chain Logistics Management* (Vol. 2, pp. 5–16). New York, NY: McGraw-Hill.
28. Bozkaya, T., & Ozsoyoglu, M., (1999). Indexing large metric spaces for similarity search queries. *ACM Transactions on Database Systems (TODS), 24*(3), 361–404.
29. Brin, S., (1995). *Near Neighbor Search in Large Metric Spaces, 1*, 1–22.
30. Brusoni, V., Console, L., Terenziani, P., & Pernici, B., (1995). Extending temporal relational databases to deal with imprecise and qualitative temporal information. In: *Recent Advances in Temporal Databases* (Vol. 1, pp. 3–22). Springer, London.
31. Buneman, P., Fan, W., & Weinstein, S., (1998). Path constraints on semistructured and structured data. In: *Proceedings of the Seventeenth ACM SIGACT-SIGMOD-SIGART Symposium on Principles of Database Systems* (Vol. 7, pp. 129–138). ACM.
32. Burkowski, F. J., (1992). An algebra for hierarchically organized text-dominated databases. *Information Processing & Management, 28*(3), 333–348.
33. Burns, J. B., & Riseman, E. M., (1992). Matching complex images to multiple 3D objects using view description networks. In: *Computer Vision and Pattern Recognition, 1992; Proceedings CVPR'92, 1992 IEEE Computer Society Conference* (pp. 328–334). IEEE.
34. Cai, J., Paige, R., & Tarjan, R., (1992). More efficient bottom-up multi-pattern matching in trees. *Theoretical Computer Science, 106*(1), 21–60.
35. Caouette, J. B., Altman, E. I., & Narayanan, P., (1998). *Managing Credit Risk: The Next Great Financial Challenge* (Vol. 2, pp. 1–20). John Wiley & Sons.
36. Cesarini, F., Lastri, M., Marinai, S., & Soda, G., (2001). Page classification for meta-data extraction from digital collections. In:

International Conference on Database and Expert Systems Applications (Vol. 1, pp. 82–91). Springer, Berlin, Heidelberg.

37. Chase, D. R., (1987). An improvement to bottom-up tree pattern matching. In: *Proceedings of the 14th ACM SIGACT-SIGPLAN Symposium on Principles of Programming Languages* (Vol. 14, pp. 168–177). ACM.
38. Chiueh, T. C., (1994). Content-based image indexing. In: *VLDB* (Vol. 94, pp. 582–593).
39. Christmas, W. J., Kittler, J., & Petrou, M., (1995). Structural matching in computer vision using probabilistic relaxation. *IEEE Transactions on Pattern Analysis and Machine Intelligence, 17*(8), 749–764.
40. Chung, C. W., Min, J. K., & Shim, K., (2002). APEX: An adaptive path index for XML data. In: *Proceedings of the 2002 ACM SIGMOD International Conference on Management of Data* (Vol. 1, pp. 121–132). ACM.
41. Ciaccia, P., Patella, M., & Zezula, P., (1997). Deis-csite-cnr. In: *Proceedings of the International Conference on Very Large Data Bases* (Vol. 23, pp. 426–435).
42. Clarke, C. L., Cormack, G. V., & Burkowski, F. J., (1995). An algebra for structured text search and a framework for its implementation. *The Computer Journal, 38*(1), 43–56.
43. Cole, R., & Hariharan, R., (1997). Tree pattern matching and subset matching in randomized O(nlog^{3}m) time. In: *Proceedings of the Twenty-Ninth Annual ACM Symposium on Theory of Computing* (Vol. 1, pp. 66–75). ACM.
44. Cole, R., Hariharan, R., & Indyk, P., (1999). Tree pattern matching and subset matching in deterministic O (n log super (3) n)-time. In: *The 1999 10th Annual ACM-SIAM Symposium on Discrete Algorithms* (Vol. 10, pp. 245–254).
45. Consens, M. P., & Mendelzon, A. O., (1990). GraphLog: A visual formalism for real life recursion. In: *Proceedings of the Ninth ACM SIGACT-SIGMOD-SIGART Symposium on Principles of Database Systems* (Vol. 1, pp. 404–416). ACM.
46. Console, L., Brusoni, V., Pernici, B., & Terenziani, P., (1995). *Extending Temporal Relational Databases to Deal with Imprecise and Qualitative Temporal Information, 1*, 1–20.

47. Conte, D., Foggia, P., Sansone, C., & Vento, M., (2004). Thirty years of graph matching in pattern recognition. *International Journal of Pattern Recognition and Artificial Intelligence, 18*(03), 265–298.
48. Cook, D. J., & Holder, L. B., (1993). Substructure discovery using minimum description length and background knowledge. *Journal of Artificial Intelligence Research, 1*, 231–255.
49. Cooper, B. F., Sample, N., Franklin, M. J., Hjaltason, G. R., & Shadmon, M., (2001). A fast index for semistructured data. In: *VLDB* (Vol. 1, pp. 341–350).
50. Corneil, D. G., & Gotlieb, C. C., (1970). An efficient algorithm for graph isomorphism. *Journal of the ACM (JACM), 17*(1), 51–64.
51. Cour, T., Srinivasan, P., & Shi, J., (2007). Balanced graph matching. In: *Advances in Neural Information Processing Systems* (Vol. 1, pp. 313–320).
52. Day, Y. F., Dagtas, S., Iino, M., Khokhar, A., & Ghafoor, A., (1995). Object-oriented conceptual modeling of video data. In: *Data Engineering, 1995; Proceedings of the Eleventh International Conference* (Vol. 1, pp. 401–408). IEEE.
53. Dehaspe, L., Toivonen, H., & King, R. D., (1998). Finding frequent substructures in chemical compounds. In: *KDD* (Vol. 98, p. 1998).
54. Dekhtyar, A., Ross, R., & Subrahmanian, V. S., (2001). Probabilistic temporal databases, I: Algebra. *ACM Transactions on Database Systems (TODS), 26*(1), 41–95.
55. Deutsch, A., Fernandez, M., Florescu, D., Levy, A., & Suciu, D., (1999). A query language for XML. *Computer Networks, 31*(11–16), 1155–1169.
56. DeWitt, D. J., Kabra, N., Luo, J., Patel, J. M., & Yu, J., (1994). Client-server paradise. In: *Proceedings of the International Conference on Very Large Databases (VLDB), 1*, 1–22.
57. Dinitz, Y., Itai, A., & Rodeh, M., (1999). On an algorithm of zemlyachenko for subtree isomorphism. *Information Processing Letters, 70*(3), 141–146.
58. Djoko, S., Cook, D. J., & Holder, L. B., (1997). An empirical study of domain knowledge and its benefits to substructure discovery. *IEEE Transactions on Knowledge and Data Engineering, 9*(4), 575–586.

59. Dolog, P., Stuckenschmidt, H., Wache, H., & Diederich, J., (2009). Relaxing RDF queries based on user and domain preferences. *Journal of Intelligent Information Systems, 33*(3), 239.
60. Drira, K., & Rodriguez, I. B., (2009). *A Demonstration of an Efficient Tool for Graph Matching and Transformation, 1,* 1–20.
61. Dubiner, M., Galil, Z., & Magen, E., (1994). Faster tree pattern matching. *Journal of the ACM (JACM), 41*(2), 205–213.
62. Dublish, P., (1990). Some comments on the subtree isomorphism problem for ordered trees. *Information Processing Letters, 36*(5), 273–275.
63. Dubois, D., & Prade, H., (1989). Processing fuzzy temporal knowledge. *IEEE Transactions on Systems, Man, and Cybernetics, 19*(4), 729–744.
64. Dutta, S., (1989). Generalized events in temporal databases. In: *Data Engineering, 1989; Proceedings Fifth International Conference* (Vol. 5, pp. 118–125). IEEE.
65. Dyreson, C. E., & Snodgrass, R. T., (1998). Supporting valid-time indeterminacy. *ACM Transactions on Database Systems (TODS), 23*(1), 1–57.
66. Eiter, T., Lukasiewicz, T., & Walter, M., (2001). A data model and algebra for probabilistic complex values. *Annals of Mathematics and Artificial Intelligence, 33*(2–4), 205–252.
67. Engels, G., Lewerentz, C., Nagl, M., Schäfer, W., & Schürr, A., (1992). Building integrated software development environments. Part I: Tool specification. *ACM Transactions on Software Engineering and Methodology (TOSEM), 1*(2), 135–167.
68. Eshera, M. A., & Fu, K. S., (1984). A graph distance measure for image analysis. *IEEE Transactions on Systems, Man, and Cybernetics*, (3), 398–408.
69. Fernández, M. L., & Valiente, G., (2001). A graph distance metric combining maximum common subgraph and minimum common supergraph. *Pattern Recognition Letters, 22*(6, 7), 753–758.
70. Fernandez, M., Florescu, D., Kang, J., Levy, A., & Suciu, D., (1998). Catching the boat with strudel: Experiences with a web-site management system. In: *ACM SIGMOD Record* (Vol. 27, No. 2, pp. 414–425). ACM.
71. Ferro, A., Gallo, G., & Giugno, R., (1999). Error-tolerant database for structured images. In: *International Conference on Advances in Visual*

Information Systems (pp. 51–59). Springer, Berlin, Heidelberg.

72. Ferro, A., Gallo, G., Giugno, R., & Pulvirenti, A., (2001). Best-match retrieval for structured images. *IEEE Transactions on Pattern Analysis and Machine Intelligence, 23*(7), 707–718.
73. Filotti, I. S., & Mayer, J. N., (1980). A polynomial-time algorithm for determining the isomorphism of graphs of fixed genus. In: *Proceedings of the Twelfth Annual ACM Symposium on Theory of Computing* (pp. 236–243). ACM.
74. Foggia, P., Sansone, C., & Vento, M., (2001). A database of graphs for isomorphism and sub-graph isomorphism benchmarking. In: *Proc. of the 3rd IAPR TC-15 International Workshop on Graph-Based Representations* (Vol. 3, pp. 176–187).
75. Fortin, S., (1996). Technical report 96–20, University of Alberta, Edomonton, Alberta, Canada. *The Graph Isomorphism Problem, 1*, 1–22.
76. Frakes, W. B., & Baeza-Yates, R., (1992). *Information Retrieval: Data Structures & Algorithms* (Vol. 331, pp. 1–30). Englewood Cliffs, New Jersey: Prentice Hall.
77. Fukagawa, D., & Akutsu, T., (2004). Fast algorithms for comparison of similar unordered trees. In: *International Symposium on Algorithms and Computation* (Vol. 1, pp. 452–463). Springer, Berlin, Heidelberg.
78. Gadia, S. K., Nair, S. S., & Poon, Y. C., (1992). Incomplete information in relational temporal databases. In: *VLDB* (Vol. 1992, pp. 395–406).
79. Ganapathysaravanabavan, G., & Warnow, T., (2001). Finding a maximum compatible tree for a bounded number of trees with bounded degree is solvable in polynomial time. In: *International Workshop on Algorithms in Bioinformatics* (pp. 156–163). Springer, Berlin, Heidelberg.
80. Garey, M. R., & Johnson, D. S., (1979). *Computers and Intractability: A Guide to NP-Completeness.*(pp. 4-7)
81. Gold, S., & Rangarajan, A., (1996). A graduated assignment algorithm for graph matching. *IEEE Transactions on Pattern Analysis and Machine Intelligence, 18*(4), 377–388.
82. Goldman, R., & Widom, J., (1997). *DataGuides: Enabling Query Formulation and Optimization in Semistructured Databases, 1*, 1–22.

83. Gonnet, G. H., & Tompa, F. W., (1987). *Mind Your Grammar: A New Approach to Modeling Text*. UW Centre for the New Oxford English Dictionary.

84. Goodman, J. E., & O'Rourke, J., (1997). *Handbook of Discrete and Computational Geometry* (Vol. 6, pp. 1–20). CRC Press series on Discrete Mathematics and its Applications.

85. Grossi, R., (1991). A note on the subtree isomorphism for ordered trees and related problems. *Information Processing Letters, 39*, 81–84.

86. Grossi, R., (1993). On finding common subtrees. *Theoretical Computer Science, 108*(2), 345–356.

87. Gupta, A., & Nishimura, N., (1995). Finding smallest supertrees. In: *International Symposium on Algorithms and Computation* (pp. 112–121). Springer, Berlin, Heidelberg.

88. Gupta, A., & Nishimura, N., (1998). Finding largest subtrees and smallest supertrees. *Algorithmica, 21*(2), 183–210.

89. Güting, R. H., (1994). GraphDB: Modeling and querying graphs in databases. In: *VLDB* (Vol. 94, pp. 12–15).

90. Gyssens, M., Paredaens, J., & Van, G. D., (1989). *A Grammar-Based Approach Towards Unifying Hierarchical Data Models* (Vol. 18, No. 2, pp. 263–272). ACM.

91. Hirata, K., & Kato, T., (1992). Query by visual example. In: *International Conference on Extending Database Technology* (Vol. 1, pp. 56–71). Springer, Berlin, Heidelberg.

92. Hirschberg, D. S., & Wong, C. K., (1976). A polynomial-time algorithm for the knapsack problem with two variables. *Journal of the ACM (JACM), 23*(1), 147–154.

93. Hlaoui, A., & Wang, S., (2002). A new algorithm for inexact graph matching. In: *Pattern Recognition, 2002; Proceedings 16th International Conference* (Vol. 4, pp. 180–183). IEEE.

94. Hlaoui, A., & Wang, S., (2004). A node-mapping-based algorithm for graph matching. *Submitted (and revised) to J. Discrete Algorithms, 1*, 1–22.

95. Hoffmann, C. M., & O'Donnell, M. J., (1982). Pattern matching in trees. *Journal of the ACM (JACM), 29*(1), 68–95.

96. Hong, P., & Huang, T. S., (2001). Spatial pattern discovering by learning the isomorphic subgraph from multiple attributed relational

graphs. *Electronic Notes in Theoretical Computer Science, 46*, 113–132.

97. Hong, P., & Huang, T. S., (2004). Spatial pattern discovery by learning a probabilistic parametric model from multiple attributed relational graphs. *Discrete Applied Mathematics, 139*(1–3), 113–135.
98. Hong, P., Wang, R., & Huang, T., (2000). Learning patterns from images by combining soft decisions and hard decisions. In: *Computer Vision and Pattern Recognition, 2000; Proceedings IEEE Conference* (Vol. 1, pp. 78–83). IEEE.
99. Hopcroft, J. E., & Wong, J. K., (1974). Linear time algorithm for isomorphism of planar graphs (preliminary report). In: *Proceedings of the Sixth Annual ACM Symposium on Theory of Computing* (Vol. 1, pp. 172–184). ACM.
100. James, C. A., Weininger, D., & Delany, J., (2000). *Daylight Theory Manual 4.71, Daylight Chemical Information Systems*. Inc., Irvine, CA.
101. Jensen, C. S., & Snodgrass, R. T., (1999). Temporal data management. *IEEE Transactions on Knowledge and Data Engineering, 11*(1), 36–44.
102. Jensen, C. S., Dyreson, C. E., Böhlen, M., Clifford, J., Elmasri, R., Gadia, S. K., & Kline, N., (1998). The consensus glossary of temporal database concepts—February 1998 version. In: *Temporal Databases: Research and Practice* (pp. 367–405). Springer, Berlin, Heidelberg.
103. Kannan, R., (1980). A polynomial algorithm for the two-variable integer programming problem. *Journal of the ACM (JACM), 27*(1), 118–122.
104. Kanza, Y., & Sagiv, Y., (2001). Flexible queries over semistructured data. In: *Proceedings of the Twentieth ACM SIGMOD-SIGACT-SIGART Symposium on Principles of Database Systems* (Vol. 1, pp. 40–51). ACM.
105. Kao, M. Y., Lam, T. W., Sung, W. K., & Ting, H. F., (1999). A decomposition theorem for maximum weight bipartite matchings with applications to evolutionary trees. In: *European Symposium on Algorithms* (Vol. 1, pp. 438–449). Springer, Berlin, Heidelberg.
106. Karray, A., Ogier, J. M., Kanoun, S., & Alimi, M. A., (2007). An ancient graphic documents indexing method based on spatial similarity. In: *International Workshop on Graphics Recognition* (Vol. 1, pp. 126–134). Springer, Berlin, Heidelberg.

107. Kato, T., Kurita, T., Otsu, N., & Hirata, K., (1992). A sketch retrieval method for full color image database-query by visual example. In: *Pattern Recognition, 1992; Conference A: Computer Vision and Applications, Proceedings, 11th IAPR International Conference* (Vol. 1, pp. 530–533). IEEE.

108. Kaushik, R., Bohannon, P., Naughton, J. F., & Korth, H. F., (2002). Covering indexes for branching path queries. In: *Proceedings of the 2002 ACM SIGMOD International Conference on Management of Data* (pp. 133–144). ACM.

109. Kilpeläinen, P., & Mannila, H., (1993). Retrieval from hierarchical texts by partial patterns. In: *Proceedings of the 16th Annual International ACM SIGIR Conference on Research and Development in Information Retrieval* (Vol. 1, pp. 214–222). ACM.

110. Kilpeläinen, P., & Mannila, H., (1994). Query primitives for tree-structured data. In: *Annual Symposium on Combinatorial Pattern Matching* (Vol. 1, pp. 213–225). Springer, Berlin, Heidelberg.

111. Kilpeläinen, P., (1992). *Tree Matching Problems with Applications to Structured Text Databases, 1*, 1–20.

112. Leordeanu, M., & Hebert, M., (2005). A spectral technique for correspondence problems using pairwise constraints. In: *Computer Vision, 2005, ICCV 2005; Tenth IEEE International Conference* (Vol. 2, pp. 1482–1489). IEEE.

113. Leordeanu, M., Hebert, M., & Sukthankar, R., (2009). An integer projected fixed point method for graph matching and map inference. In: *Advances in Neural Information Processing Systems* (pp. 1114–1122).

114. Luccio, F., & Pagli, L., (1991). Simple solutions for approximate tree matching problems. In: *Colloquium on Trees in Algebra and Programming* (pp. 193–201). Springer, Berlin, Heidelberg.

115. Lueker, G. S., & Booth, K. S., (1979). A linear time algorithm for deciding interval graph isomorphism. *Journal of the ACM (JACM), 26*(2), 183–195.

116. Luks, E. M., (1982). Isomorphism of graphs of bounded valence can be tested in polynomial time. *Journal of Computer and System Sciences, 25*(1), 42–65.

117. Luo, B., & Hancock, E. R., (2001). Structural graph matching using the EM algorithm and singular value decomposition. *IEEE Transactions on Pattern Analysis and Machine Intelligence, 23*(10), 1120–1136.

118. Macleod, I. A., (1991). A query language for retrieving information from hierarchic text structures. *The Computer Journal, 34*(3), 254–264.
119. McHugh, J., Abiteboul, S., Goldman, R., Quass, D., & Widom, J., (1997). Lore: A database management system for semistructured data. *SIGMOD Record, 26*(3), 54–66.
120. Milo, T., & Suciu, D., (1999). Index structures for path expressions. In: *International Conference on Database Theory* (pp. 277–295). Springer, Berlin, Heidelberg.
121. Navarro, G., & Baeza-yates, R., (1995). Expressive power of a new model for structured text databases. In: *In Proc. PANEL'95*, 1–20.
122. Nishimura, N., Ragde, P., & Thilikos, D. M., (2000). Finding smallest supertrees under minor containment. *International Journal of Foundations of Computer Science, 11*(03), 445–465.
123. Pelegri-Llopart, E., & Graham, S. L., (1988). Optimal code generation for expression trees: An application BURS theory. In: *Proceedings of the 15th ACM SIGPLAN-SIGACT Symposium on Principles of Programming Languages* (pp. 294–308). ACM.
124. Salminen, A., & Tompa, F. W., (1992). PAT expressions: An algebra for text search. *Acta Linguistica Hungarica, 41*(1/4), 277–306.
125. Schlieder, T., (2002). Schema-driven evaluation of approximate tree-pattern queries. In: *International Conference on Extending Database Technology* (Vol. 1, pp. 514–532). Springer, Berlin, Heidelberg.
126. Shaw, S., Vermeulen, A. F., Gupta, A., & Kjerrumgaard, D., (2016). Querying semi-structured data. In: *Practical Hive* (pp. 115–131). Apress, Berkeley, CA.
127. Sikora, F., (2012). *An (Almost Complete) State of the Art Around the Graph Motif Problem* (Vol. 1, pp. 1–22). Université Paris-Est Technical Reports.
128. Tague, J., Salminen, A., & McClellan, C., (1991). Complete formal model for information retrieval systems. In: *Proceedings of the 14th Annual International ACM SIGIR Conference on Research and Development in Information Retrieval* (pp. 14–20). ACM.
129. Umeyama, S., (1988). An eigendecomposition approach to weighted graph matching problems. *IEEE Transactions on Pattern Analysis and Machine Intelligence, 10*(5), 695–703.

130. Verma, R. M., & Reyner, S. W., (1989). An analysis of a good algorithm for the subtree problem, corrected. *SIAM Journal on Computing, 18*(5), 906–908.

131. Wilson, R. C., & Hancock, E. R., (1997). Structural matching by discrete relaxation. *IEEE Transactions on Pattern Analysis and Machine Intelligence, 19*(6), 634–648.

132. Xu, Y., Saber, E., & Tekalp, A. M., (1999). Hierarchical content description and object formation by learning. In: *Content-Based Access of Image and Video Libraries, 1999 (CBAIVL'99) Proceedings IEEE Workshop* (pp. 84–88). IEEE.

CHAPTER 4

ALGORITHMIC SEARCH VIA QUANTUM WALK

CONTENTS

4.1. INTRODUCTION

Uncategorized databases require a long time to locate a distinctive element since they are not organized. Whenever the total number of factors in the database rises, the search time increases in direct proportion to the size of the database N. It is assumed that there would be no viable search method available in the normal area of computing. The conception of a quantum computer may provide us with the promise of enhancing the answer to the searching challenges, as well as to a few other vexing problems like discrete logarithms and factoring huge numbers, which have been a source of consternation for decades (Aharonov et al., 1993; Ambainis, 2003, 2007). Grover discovers the first quantum algorithm for exploring an uncategorized database, known as the Grover algorithm. Grover algorithm is named after Grover, who discovered the algorithm (Grover, 1997; Ambains et al., 2005). Grover method is capable of searching an unsorted database at a quadratic speed up, and it has been proved to be the fastest speed increase by a quantum computer in a short period. Although, in contrast to some other quantum algorithms, the Grover algorithm exhibits a quadratic speed increase over the analogous classical algorithms, this should not prevent us from considering the Grover algorithm as elegant produce of the quantum computer, as the quadratic speed increase is not a small accomplishment when dealing with very large unsorted databases. I think a typical search involves 106 steps, with each step costing 8.64 seconds. At that point, the overall calculation will take 100 days to complete. On the other hand, the comparable quantum search requires just 103 steps, and hence the entire calculation takes just 0.1 days!

As an alternative to the Grover technique, the quantum walk algorithm may get quadratic speed increase results. Furthermore, with little effort, the quantum walk may be easily extended to other types of searching problems, such as element distinctness and substructure discovery. It is impossible to efficiently tackle these issues using the Grover (1997) technique. This is a long way from being finished. As we can see, quantum computers are still a magnificent fantasy, which has prompted various researchers to double their efforts to uncover the search algorithm dependent on the conservative quantum walk or quantum computer. A few individuals may be skeptical that decoherence may bring about the demise of the ambition of building a quantum computer. However, such cynicism will be dispelled when it is realized that quantum walk may play a key part in various natural processes, like photosynthesis. It should come as no surprise that the fantasy will one day come true (Blankenship, 2002; Nielsen and Chuang, 2002).

This chapter will concentrate on a search algorithm that depends on the quantum walk as its central component. The following is the organization of this chapter: Section two contains an outline of the quantum walk, which includes the continuous and coined quantum walks; Section three contains the search algorithm based on the quantum walk, which applies to both common and special searching difficulties; Section four is devoted to the physical application of the quantum-walk-based search algorithm via an NMR quantum computer, and Section five contains an introduction to the implementation of quantum walk in nature for instance. Finally, part six contains the conclusion and recommendations for additional research (Childs and Goldstone, 2004; Broom et al., 2010).

4.2. QUANTUM WALK

Random walk is a Markovian chain used in algorithms in general, and it was quantal once the idea of a quantum computer was developed. We will present the notion of the quantum walk, which is derived from the traditional random walk, and the two types of quantum walk: continuous and coined quantum walks, in this section (Du et al., 2003; Douglas and Wang, 2009).

4.2.1. Classical Random Walk

In theory, the random walk may be defined on any graph, regardless of its complexity. We can focus on the random walk on a single-dimensional lattice without losing sight of the generalization. For example, in Figure 4.1, the lattice points are ordered on a line and labeled with an integer, either positive or negative. The lattices are labeled from –9 to 9. We may be at a single position in the lattice at any time. Then, starting at lattice 0, for instance, we flip a coin to determine whether we will go to the right lattice or the left lattice; if the coin is up, we will go to the left lattice; or else, we will go to the right lattice; and from there, we will flip the coin again to determine whether we will go to the right or the left lattice (Farhi and Gutmann, 1998; Engel et al., 2007). At the end of every step, we flip a coin to choose which way we should move. The probability of every lattice may be determined after the T steps. This time, we set the probability bounds for every direction to be 0.5. We have the option of making the probability different if this is essential (Figure 4.1) (Kendon et al., 2003, 2007).

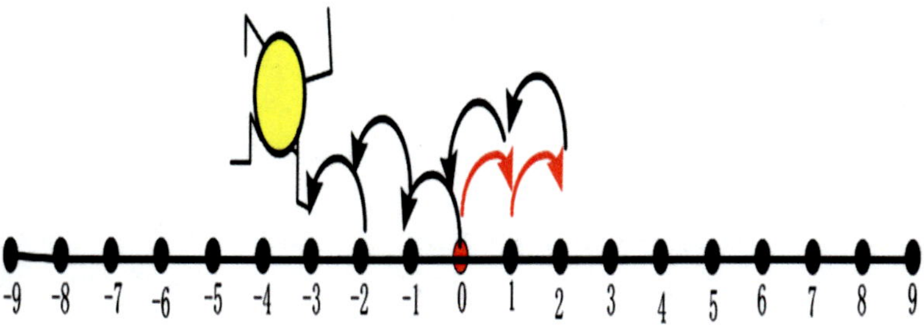

Figure 4.1. In the case of the one-dimensional lattice, a walker may only pick between two directions. In a traditional random walk, the decision to travel left or right is made by flipping a two-sided coin.

Source: https://www.intechopen.com/books/search-algorithms-and-applications/search-via-quantum-walk.

We can obtain the probabilistic model of the walker taking place on every lattice using a correct deduction. For details, we can look up the book 'A Modern Course in Statistical Physics' (Table 4.1 and Figure 4.2) (Reichl, 1998).

Table 4.1. In this table Tis the number of steps of the classical random walk in one dimensional lattice, i is the number that labels the position of the lattice. From this table we can know that after T steps, the walker will be at the center (or the start place) with the maximum probability

T \ i	-5	-4	-3	-2	-1	0	1	2	3	4	5
0						1					
1					$\frac{1}{2}$		$\frac{1}{2}$				
2				$\frac{1}{4}$		$\frac{1}{2}$		$\frac{1}{4}$			
3			$\frac{1}{8}$		$\frac{3}{8}$		$\frac{3}{8}$		$\frac{1}{8}$		
4		$\frac{1}{16}$		$\frac{1}{4}$		$\frac{3}{8}$		$\frac{1}{4}$		$\frac{1}{16}$	
5	$\frac{1}{32}$		$\frac{5}{32}$		$\frac{5}{16}$		$\frac{5}{16}$		$\frac{5}{32}$		$\frac{1}{32}$

*We may tell from this table that once *T* steps, the walker will be at the center with the highest likelihood of arriving there.

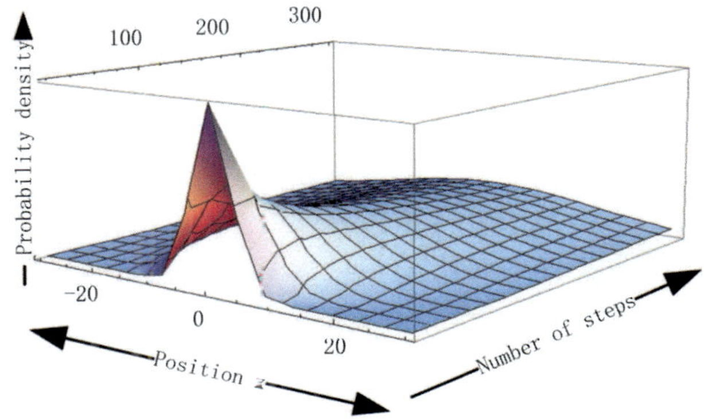

Figure 4.2. Displays a probability distribution for the classical random walk about the position and number of steps taken, demonstrating that as the number of steps rises, the walker will disperse to all lattice points. A large number of computer algorithms employ this character.

Source: https://www.intechopen.com/books/search-algorithms-and-applications/search-via-quantum-walk.

Following the probability principle, the probability distribution of any walker on a given place when sufficiently enough time has passed is:

$$\rho(x,T) = \frac{1}{\sqrt{2\pi T}} \exp\left(-\frac{x^2}{2T}\right) \quad (1)$$

where; x is the location on the one-dimensional lattice, the step number is designated by T.

It depicts the probability concentrations of the distribution as a function of the step number and the location of the distribution. Figure 4.3 depicts the results of some common procedures. It is not straightforward to discover that, after some steps, the probability of the random walker's placement on the lattice becomes even, indicating that the probability is closer to a Gauss distribution. The average position is zero, while the variance of the position is 100%.

$\sigma^2 = T$ (2)

So statistically, the walker's exit from the center is proportionate to the square root of the number of steps.

4.2.2. Coined Quantum Walk

The classic random walk has been used in various domains, including randomized algorithms and Brownian motion. However, we anticipate quantizing random walk to gain more influential applications, which is mostly due to the quantum realm's superposition principle (Figures 4.3 and 4.4).

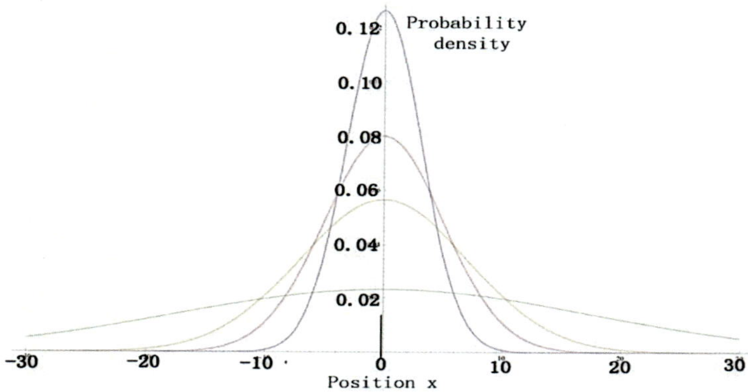

Figure 4.3. Following a series of specific steps, the probability distribution of the classical random walk is shown.

Source: https://www.intechopen.com/books/search-algorithms-and-applications/search-via-quantum-walk.

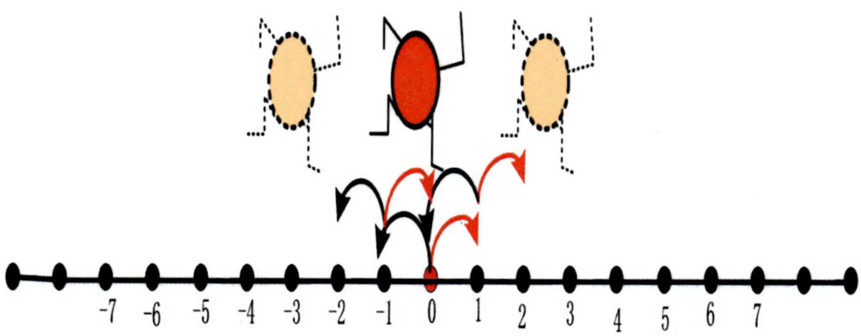

Figure 4.4. *Design of the quantum walk using intuitionistic principles. The walker may go both lefts and right at the same time in this situation, which is one of the most astonishing aspects of quantum mechanics, which was initially demonstrated by Feynman using the integral path tool (Kendon et al., 2007).*

Source: https://www.intechopen.com/books/search-algorithms-and-applications/search-via-quantum-walk.

Intuitively, the quantum walks maybe a quantization of the classical random walk. Nevertheless, in the typical random walk, the walker may only visit one lattice at a time. On the other hand, the quantum walker can turn both ways until it is measured.

We should talk about quantum walking in Hilbert space in a formal sense. We will concentrate on the single-dimensional lattice for simplicity.

We must define two Hilbert spaces in the coined quantum walk:

$$H = H_p \otimes H_c \tag{3}$$

where; H_p is the position space; and H_c is the coin Hilbert space with the following forms:

$$H_p = \{|x>; x \in \mathbb{Z}\}, H_c = \{|+1>, |-1\} \tag{4}$$

Where the integer x is the location, in the coin space $+1$ indicates to go to right; and -1 indicates to go to left. In quantum walk, the walking procedure can be comprehended by the shift operator:

$$S = |+1\rangle\langle+1| \otimes \sum_x |x+1\rangle\langle x| + |-1\rangle\left\langle -1| \otimes \sum_x |x+1\rangle\langle x\right| \tag{5}$$

The coin operator is:

$$C = \begin{pmatrix} a & b \\ c & d \end{pmatrix} \tag{6}$$

The C and S operators are all Hermitian and unitary, so for every step, the progress of the space and coin is likewise unitary as follows:

$$U = S(C \otimes I) \tag{7}$$

The state of the quantum coin could be a principle of superposition of down and up, which is unlike the conventional case. Thus, the initial situation in the quantum walk can be thought as:

$$|\Psi_{in}\rangle = (\alpha|+1\rangle + \beta|-1\rangle) \otimes |x\rangle \tag{8}$$

After T steps, the ultimate state before the measurement is:

$$U^T |\Psi_{in}\rangle \tag{9}$$

Then we may measure the walker's location and obtain the position distribution using the quantum mechanical rule.

We can set two quantum registers: the position register and the coin register. The coin register is likely to be in one of two states: |+1> or |–1>, and the place register might be in the state |x>, where x is an integer.

The walking technique is to first flip the coin and then shift, and the coin operator can be adjusted as:

$$C = \frac{1}{\sqrt{2}}\begin{pmatrix} 1 & 1 \\ 1 & -1 \end{pmatrix} \tag{10}$$

Flipping takes the lead to the following variations of the states:

$$|+1> \rightarrow (|+1> + |-1>)/\sqrt{2}, |-1> \rightarrow (|+1> - |-1>)/\sqrt{2}$$

Shift is:

$$|+1>|x> \rightarrow |+1>|x+1>, |-1>|x> \rightarrow |-1>|x-1>$$

The integral route process can also help us realize why the quantum walk differs from the classical random walk. The dotted line in figure 4.4 represents all conceivable courses, whereas the actual line represents any. The classical walker can only travel down one of the pathways, but the quantum walker may walk down all of the paths simultaneously, with each path's probability amplitude interacting with the others (Kempe, 2003; Hilley et al., 2009).

If the primary state of the coin is |–1›, we can get the probability distribution as Figure 4.4 shows. Else, suppose we set the coin's main state to 1/2 (|+1 + I |–1). In that case, the probability distribution looks like figure4.4. This differs from a traditional random walk, in which the probability distribution is unaffected by the initial conditions (Khaneja et al., 2005; Kendon, 2006; Kendon and Maloyer, 2007).

Another distinction between a classical random walk and a quantum walk is the walker's distribution rate as a function of the center. As we can see from the previous sections, the walker's deviance is relative to the root of the number of steps N; however, in quantum walking, the walker's deviance is proportional to N, resulting in a quadric speed increase.

4.2.3. Continuous-Time Quantum Walk

Sam Gutmann and Edward Farhi were the first to propose the concept of continuous-time quantum walk (1998). The distinction between a continuous quantum walks and a coined quantum walk is that a continuous quantum walk does not require a coin, and it is frequently characterized using the technique of graphs instead (Karski et al., 2009; Lu et al., 2010).

The lattice is shown by the red point in the diagram below, and the dotted line depicts all of the potential paths. In a conventional random walk, the walker can just select one of the pathways with a probability; however, in a quantum walk, the walker can travel along any or all of the paths simultaneously, regardless of their likelihood. If this is permitted, the probability amplitudes of each path will be able to interfere with one another (Figures 4.5 and 4.6) (Pérez Delgado, 2007; Mohseni et al., 2008).

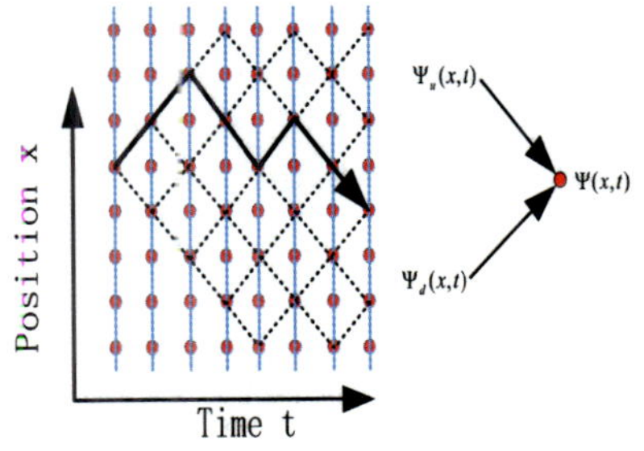

Figure 4.5. With the help of the integral route tool, we can visualize the quantum walk (Pérez Delgado, 2007).

Source: https://www.intechopen.com/books/search-algorithms-and-applications/search-via-quantum-walk.

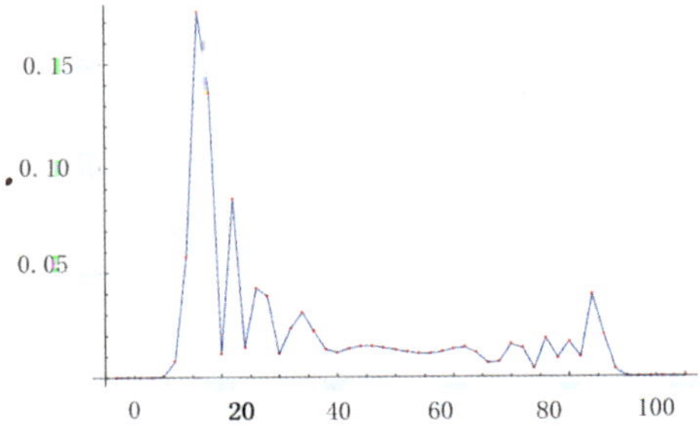

Figure 4.6. The probability distribution of a quantum walks given the coin's initial state |–1>. The walker begins at x = 50, and the total number of steps is 50.

Source: https://www.intechopen.com/books/search-algorithms-and-applications/search-via-quantum-walk.

The walker begins at x=50, and the total number of steps taken is 50. The classical random walk from a graph may describe the continuous-time quantum beginning. A matrix M that alters the possibility distribution on the graph vertex can be used to explain the procedure:

$$p_i^{t+1} = \sum_j M_{ij} p_j^t \tag{11}$$

where; M_{ij} is the matrix component of *M*, the probability on the ith vertex at time *t*.

To make the transformation process continuous, the next step is to leap to a vertex adjacent to the one that was transformed. Afterward, we need to familiarize ourselves with the H (infinitesimal generator matrix), which is used to explain the walking process, i.e.

$$\frac{dp_i(t)}{dt} = -\sum_j H_{ij} p_j(t) \tag{12}$$

It is simple to see that the possibility distribution of quantum walk is dependent on the primary state of the coin when we compare Figures 4.7 and 4.8. This is in contrast to the possibility distribution of classical random walk, which is separate from the primary state of the coin when we compare Figures 4.7 and 4.8.

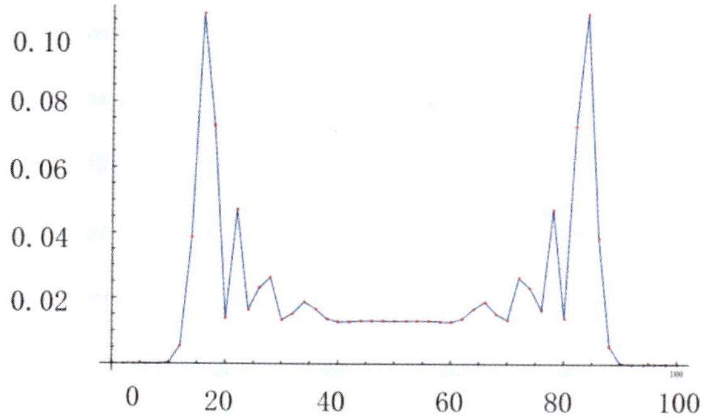

Figure 4.7. Probability distribution of quantum walks with another 1/√2 (| +1 + i | −1) starting state of the coin. At x = 50, the walker starts.

Source: https://www.intechopen.com/books/search-algorithms-and-applications/search-via-quantum-walk.

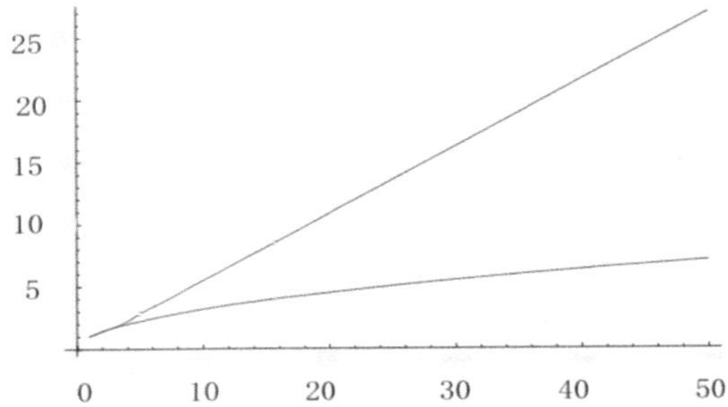

Figure 4.8. The walker's diffusion rate from the center is defined as the divergence of the walker's location from the center. The number of steps T is represented on the horizontal axis, while the divergence of the position is represented on the vertical axis.

Source: https://www.intechopen.com/books/search-algorithms-and-applications/search-via-quantum-walk.

In the diagram above, the blue line represents a quantum walk, whereas the red line represents a conventional random walk. It can be demonstrated that the quantum walk's diffusion rate stays quadratic compared to the classical random walk.

Solving the equation, we acquire:

$$p(t) = \exp(-Ht)p(0) \qquad (13)$$

where; p is the vector of possibility distribution.

Eqn. (12) has a similar form with the Schrödinger equation; thus, the classical random walk can be quantized in a continuous form (Perets et al., 2008; Potocek et al., 2009).

4.3. SEARCH ALGORITHM VIA QUANTUM WALK

It has been established in earlier sections that in a single-dimensional pattern, the walker progresses from the center quadratically faster than in a classical random walk. However, it is not a search process. When you search, you begin from a continual superposition of all the files and return to the object you previously identified as being searched for. As a result, it is not hard to see why the quantum walk-centered search method represents, on average, a quadratic performance improvement over the conventional search process. The generalized quantum walk has the advantage of turning quantum walk more quickly than the continuous-time quantum walks (Ryan et al., 2005, 2008; Panitchayangkoon et al., 2010).

The quantum walk may also be reproduced by a quantum circuit, which can theoretically implement the quantum-focused search method using a quantum computer (Douglas and Wang, 2009). This distinguishes quantum walk as not merely a hypothetical tool for algorithm development but also a valuable contribution to computing theory in the pursuit of more advanced algorithms for intractable problems. We may look at numerous publications on the present issue to better understand the algorithm implementation of quantum walk (Rebentrost et al., 2009; Reitzner et al., 2009).

4.3.1. Searching on Graphs

In this section, we will stress graphs to conduct our searches. The goal of searching on graphs is to locate a marked vertex on the graph that has been marked. In addition to finding a marked sub-graph or a marked edge, which is a generalization of the search for vertex, one may also be concerned with discovering a marked sub-graph (Shue and Zamani, 1993; Suryaprakash, 2000; Schmitz et al., 2009).

Searching for nodes on the graph may be accomplished using either continuous-time quantum walk or coined quantum walk techniques. The dubbed quantum walk-focused search method will be the primary topic of this section. Previously, all quantum walk explanations were based on very symmetric networks, such as the hypercube (Shue et al., 2001; Tulsi, 2008; Schreiber et al., 2010).

The sole difference between a quantum walks on an upper-dimensional graph and a quantum walk on a line is the Hilbert space length of the coin and its location in the graph. The coin's position is substituted with the vertex in the graph, and the Hilbert space dimension of the coin remains the degree of the graph (Zähringer et al., 2010; Simard and L'Ecuyer, 2011). For n hypercube having n dimensions, the degree of each vertex is n and the total number of the nodes is $N = 2^n$. Hence the Hilbert space of the coin and vertex is:

$$H = H_c \otimes H_v \qquad (14)$$

where; H_c is the coin space; and H_v is the vertex space which have the form:

$$H_v = \{|x\rangle : x \in \mathbb{Z}_N\} \qquad (15)$$

$$H_c = \{|c\rangle : c \in \mathbb{Z}_d\} \qquad (16)$$

where; d is the degree of each vertex; and N is the total number of the nodes. Then we may define the shift operator and coin operator by the Hilbert space of the vertex and coin in the following forms:

$$S = \sum_{d=0}^{n-1} \sum_{\vec{x}} \left| d, \vec{x} \otimes \vec{e_d} \right\rangle \left\langle d, \vec{x} \right| \tag{17}$$

$$C = I \otimes C_0 \tag{18}$$

where; e_d is the hypercube's d^th root vector; C_0 is the n n unitary operator for the coin space; while I is the identity operator. A marked coin operator is commonly used to apply the search algorithm. The marked coin operator in the SKW algorithm, for example, is as follows:

$$C' = C_0 \otimes I + (C_1 - C_0) \otimes \left|\vec{0}\right\rangle \left\langle \vec{0} \right| \tag{19}$$

The marked coin might be an x n unitary operator; we can look up many studies on the subject for further detail. An edge or a sub-graph might represent a more generic search aim on the graph. Element clearness is another technique that may be considered a quantum walk-based search (Shi and Eberhart, 1998, 1999; Secrest and Lamont, 2003).

4.3.2. Searching the Exits

Andrew M. Childs and colleagues developed a new search method based on the quantum walk concept (2003), the first algorithm to be centered on the quantum walk. Based on a classical random walk, this technique exploits the exponential speedup of the striking time of the quantum walk. In contrast to the unsorted database, the technique developed by Childs and colleagues (2003) is focused on a specific form of network. Assuming we are at the network's entrance, we aim to locate another departure point as quickly as feasible. The most efficient strategy may be to pick the direction freely to solve a classical problem, as in the classical random walk method. It continues to take time, with the length of the network increasing exponentially over time. It is possible to become disoriented at the core of the network. In quantum walk, on the other hand, one may take all of the possible pathways at the same time and arrive at the exit site with the time rising polynomially with the size of the system, resulting in an exponential increase in speed (Figure 4.9) (Sadikov and Bratko, 2006; Sanders and Kandrot, 2010).

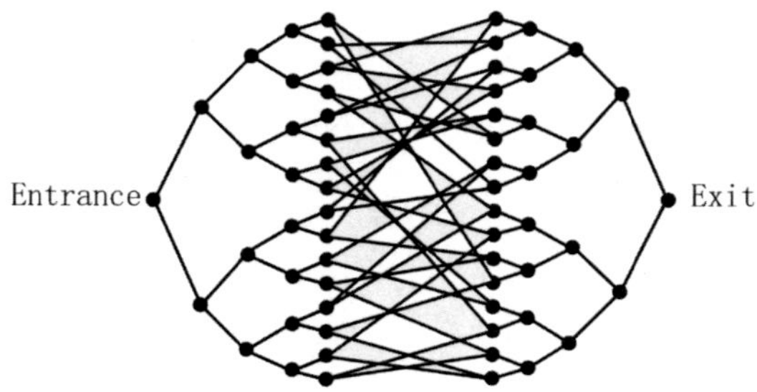

Figure 4.9. The first algorithm's network, based on a quantum walk, has been discovered (Childs et al., 2003).

Source: https://www.intechopen.com/books/search-algorithms-and-applications/search-via-quantum-walk.

4.4. THE PHYSICAL IMPLEMENTATION OF QUANTUM WALK BASED SEARCH

4.4.1. Introduction to the SKW Algorithm

Shenvi et al. (2003) proposed a random quantum walk search algorithm as one of the novel algorithms demonstrating quantum computation's supremacy. It's a quantum random walk model with independent time. SKW algorithm, like Grover's quantum search method, executes a revelation search on a database of N items with O calls, where N is the search space's scope. However, even if the Grover method's diffusion phase cannot be performed efficiently, this approach may still be used, which is a big benefit of the Grover algorithm (Ambainis, 2005; Chandrashekar, 2008; Tulsi, 2008; Reitzner et al., 2009). Many SKW algorithm improvements were proposed to reduce the method's complexity.

The primary problem can be defined as follows: specified a function $f(x), f(x) = 1$ if $x = a$, else $f(x) = 0$. The aim is to discover a, where $0\ a\ 2^n - 1$. It is comparable to searching for the only marked node amongst the $N = 2^n$ nodes in the n-cube.

$$\frac{1}{2} - O(1/n)$$

To perform the coined quantum, walk model, a coin must be flipped. The model provides a two-step approach that includes a coin-flip step and a coin-controlled walk step. In the equation U = SC, C indicates a unitary operation conforming to flipping the quantum coin. S is a variation matrix that conducts a controlled shift centered on the current state of coin space (step of coin-controlled walking). Particularly in the case of the SKW algorithm, an oracle is necessary to comprehend the search operation. The oracle functions by establishing a connection between a marking coin (C_1) and the marked node and a connection between the original coin (C_0) and the unmarked nodes, which is specified as a new coin operator C. After relating $U' = SC'$ for t_f^n times, we get the marked node with probability through measurement (Ryoo et al., 2008; Rost et al., 2009; Roberge and Tarbouchi, 2012).

For n = 2, a basic example is studied. We'll require three qubits to describe the algorithm: one coin qubit and two database qubits. As defined by the 1-out-of-4 method, any of the four computational bases is the target node. Now we'll go through the search technique in more depth. Assume the main state is one of purity |000>.

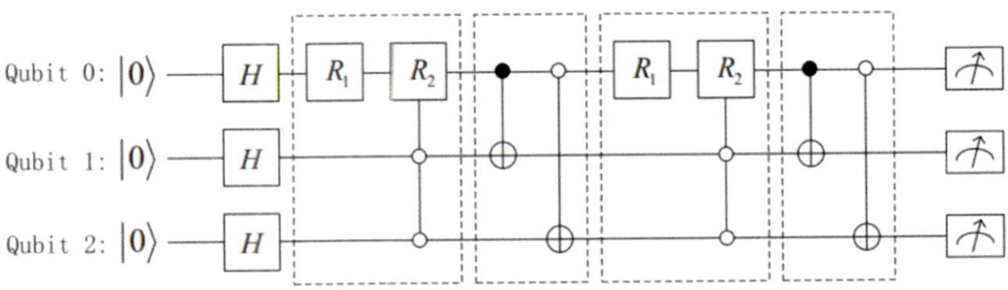

The target state of the quantum network for the method 1-out-of-4 searches is |00>12. Qubit 0 is a coin qubit, but qubits 1 and 2 are database qubits. To generate an equal superposition overall computing bases, Hadamard gates are used. The solid circle represents a single control gate, whereas the open circle represents the inverse. When the database is |00 >12 and $C_0 = R^0x(3/2)$ otherwise, the goal of oracle C ′ is to apply $C_1 = R^0x$ (/2) when the database is |00 >12. It's comparable to substituting $R_1 = R^0x$ (3/2) and $R_2 = R^0x$ (). If qubit 0 is |1>0, the two controlled-not gates invert qubit one and qubit two if qubit 0 is |0>0,' respectively. The measurement necessitates the reconstruction of all populations. For different goal states, similar circuits may be created easily. For example, if the goal is |10>12, all we have to do is change the regulated condition of the three-body-interface gate to state |10>12.

I. Employing a Hadamard operation to each qubit to formulate the state:

$$\left|\psi_i\right\rangle = \frac{\left|0\right\rangle_0 + \left|1\right\rangle_0}{\sqrt{2}} \otimes \frac{\left|0\right\rangle_1 + \left|1\right\rangle_1}{\sqrt{2}} \otimes \frac{\left|0\right\rangle_2 + \left|1\right\rangle_2}{\sqrt{2}} \quad (20)$$

Which is precisely an equal superposition across all the computational bases.

II. Execute the vision C' on the qubit of coin contingent with the state of database qubits, i.e., $C_1 = R^0_x(\pi/2) = e^{-i\pi\sigma x/4}$ if the qubits of the database are on the target state $\left|\tau\sigma\right\rangle_{12}$, and $C_0 = R^0_x(3\pi/2) = e^{-i3\pi\sigma x/4}$ else. Thus, the entire coin operation is:

$$C' = C_0 \otimes (E_{12} - \left|\tau\sigma\right\rangle_{1212}\left\langle\tau\sigma\right|) + C_1 \otimes \left|\tau\sigma\right\rangle_{1212}\left\langle\tau\sigma\right| \quad (21)$$

where; E_{12} is the character operator. Then the database qubits experience the shift operation S accustomed on the state of qubit of the coin:

$$\begin{aligned} &\left|0\right\rangle_0\left|00\right\rangle_{12} \Leftrightarrow \left|0\right\rangle_0\left|01\right\rangle_{12} \\ &\left|0\right\rangle_0\left|10\right\rangle_{12} \Leftrightarrow \left|0\right\rangle_0\left|11\right\rangle_{12} \\ &\left|1\right\rangle_0\left|00\right\rangle_{12} \Leftrightarrow \left|1\right\rangle_0\left|01\right\rangle_{12} \\ &\left|1\right\rangle_0\left|01\right\rangle_{12} \Leftrightarrow \left|1\right\rangle_0\left|11\right\rangle_{12} \end{aligned} \quad (22)$$

III. Repeat step (II) two times to implement the quantum walk that will reach the final state:

$$\left|\psi_f\right\rangle = (SC')^2\left|\psi_i\right\rangle \quad (23)$$

IV. Measure all the populaces of the database qubits. When looking for |00>12, for example, we can see that the possibilities of |00>12, |01>12, |10>12, and |11>12 are 0.5, 0.25, 0.25, and 0, respectively. Analogous networks may readily be created for various goal states with the controlled condition altered to the target node. The outcomes are comparable to those of the previous study (Luštrek, 2005; Luštrek and Bulitko, 2006; Lindholm et al., 2008).

4.4.2. NMR Experimental Implementation

We will now use the NMR quantum computer to execute the SKW algorithm on the data. In a sample of 1-bromo-2,3-dichlorobenzene organized in a liquid-crystal solvent, the three (1H) spins are designated by the three (1H) spins in the sample (ZLI-1132). The following equation depicts the molecular structure of the compound (a). The system Hamiltonian can be defined as:

$$H = \sum_{j=1}^{3} 2\pi v_j I_z^j + \sum_{j,k,j<k\le 3} 2\pi J_{jk}\left(I_x^j I_x^k + I_y^j I_y^k + I_z^j I_z^k\right) + \sum_{j,k,j<k\le 3} 2\pi D_{jk}\left(2I_z^j I_z^k - I_x^j I_x^k - I_y^j I_y^k +\right) \tag{24}$$

where, J_{jk} and D_{ijk} are the dipolar coupling strengths and scalar coupling strengths, respectively, among the kth and j[th] spins, and v_j is the resonance frequency of the jth spin. All amounts are confined to the spins contained within a single molecule alone. The author carried all investigations at room temperature on a Bruker Avance (500 MHz) spectrometer. The equation depicts the spectrum of the thermal equilibrium state after being tracked by a/2 hard pulse for some time (b). We iteratively link the estimated and observed spectra to the constraints' disconcertion, starting with the originally assumed parameters presuming the molecule geometry and working our way up. Table 4.2 contains the values of the parameters. Because the system Hamiltonian contains non-diagonal components, the eigenstates are no longer regarded as Zeeman product states but rather as linear combinations of the other states (Figure 4.10 and Table 4.2).

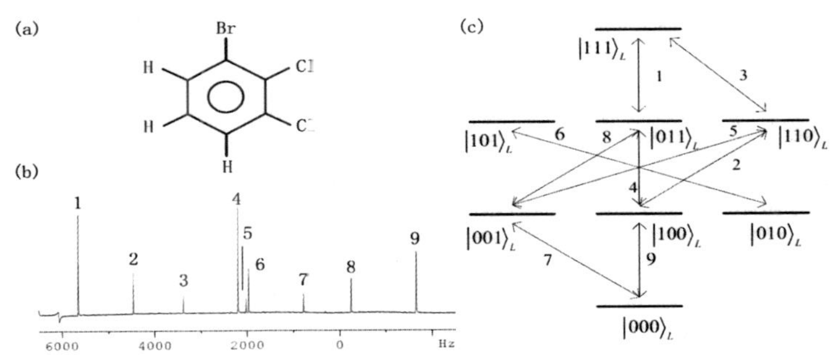

Figure 4.10. (a) The three protons create a 3-qubit system in the molecule of 1-bromo-2,3-dichlorobenzene. (b) After the π/2 hard pulse, the spectrum of the thermal equilibrium state is shown. The ordering of the frequencies is used to name all observable transitions. (c) In the eigenbasis, a diagram of the associated transitions is shown.

Source: https://www.intechopen.com/books/search-algorithms-and-applications/search-via-quantum-walk.

Table 4.2. (a) The Fitting Parameters for 1-Bromo-2,3-Dichlorobenzene's Spectrum (Hertz)* (b) The findings are displayed on the No. 9, No. 8, and No. 7 transitions.

(a)	H_1	H_2	H_3
H_3	2147.2	1.4	8
H_2	–339.35	2094.8	8
H_1	–1341.7	1633.4	1945.5

(b)

spin 1 No.9	$UR_y^1(\pi/2)$	$UR_y^1(\pi/2)R_y^2(\pi)$	$UR_y^1(\pi/2)R_y^3(\pi)$	$UR_y^1(\pi/2)R_y^{2,3}(\pi)$
	$P(1)-P(5)$	$P(3)-P(7)$	$P(2)-P(6)$	$P(4)-P(8)$
spin 2 No.8	$UR_y^2(\pi/2)$	$UR_y^2(\pi/2)R_y^1(\pi)$	$UR_y^2(\pi/2)R_y^3(\pi)$	$UR_y^2(\pi/2)R_y^{1,3}(\pi)$
	$P(2)-P(4)$	$P(6)-P(8)$	$P(1)-P(3)$	$P(5)-P(7)$
spin 3 No.7	$UR_y^3(\pi/2)$	$UR_y^3(\pi/2)R_y^1(\pi)$	$UR_y^3(\pi/2)R_y^2(\pi)$	$UR_y^3(\pi/2)R_y^{1,2}(\pi)$
	$P(1)-P(2)$	$P(5)-P(6)$	$P(3)-P(4)$	$P(7)-P(8)$

*The diagonal elements represent the chemical shifts of the three protons, the upper-right off-diagonal elements represent dipolar coupling strengths, and the lower-left ones represent scalar coupling strengths.

We constructed a U to implement the conversion about the computational eigenbasis and basis, which fulfills to simplify the size of populations (labeled from P(1) to P(8)).

$$H_L = UH_SU^{\dagger} \quad (25)$$

The system Hamiltonian is (HS) while the diagonal Hamiltonian is (HL) (that is, the Hamiltonian in the eigenbasis). We may immediately obtain all eight population values by calculating the pulse of applying transformation matrix (U) after the initial read-out pulses in liquid NMR and merging with the normalization 1. Table 4.2(b) shows all of the P(i) P(j) values that may be obtained using various read-out pulses (Liang and Suganthan, 2005; Liang et al., 2006; Li et al., 2007).

$$\rho_{000} = \frac{1-\varepsilon}{8}1 + \varepsilon|000\rangle\langle000|$$

The experiment was separated into three phases: The pseudo-pure state preparation and population measurement are all part of the quantum random walk searching process. We must first produce the PPS from the thermal equilibrium state. To understand the PPS preparation, we employed shape pulses centered on the gradient ascent pulse engineering (GRAPE) technique and gradient pulses, with a simulated numerical fidelity of 0.977 (L'Ecuyer and Simard, 2007; Koenig and Sun, 2009).

The quantum random walk searching process is divided into two parts: initial state preparation and two iterations of unitary development. We combined them into a GRAPE pulse of 20 ms and 250 sections with fidelity of more than 0.990. When creating GRAPE pulses of 20 ms with fidelity 0.990, the reading-out operations listed in Table 4.2(b) are also used. The probability of achieving |00>12, |, |10>12, and |11>12 is 0.513, 0.232, 0.197, and 0.058, respectively, indicating that we have completed our SKW algorithm-centered search (Koenig, 2004; Khan et al., 2012; Kislitsyn et al., 2017).

Besides $|00\rangle_{12}$, we altered the objective states to $|01\rangle_{12}$, $|10\rangle_{12}$ and $|11\rangle_{12}$. On the graph, you can see how the experiments turned out. Both theoretical and experimental results are largely consistent, with just a small variation. The tiny divergence between theory and experiment can be ascribed to decoherence, inhomogeneity of the RF field, and improper implementation of GRAPE pulses, among other factors. We utilized GRAPE pulses to comprehend high-fidelity unitary operations, and we developed a method for measuring the populations of the density matrix that was both efficient and accurate. The experimental results are extremely close to those predicted by the theory, demonstrating the superiority of the algorithm in this case (Kennedy, 1999, 2003, 2011).

4.5. QUANTUM WALK-BASED SEARCH IN NATURE

One of the most amazing miracles of nature is photosynthesis, which produces all of the chemical energy required by the planet while also serving as an essential source of energy storage. Despite this, the high efficiency of energy transmission in photosynthesis remains a mystery. To understand the mechanism of energy transfer, the method of the quantum walk has been familiarized, as the quantum walk can increase the search efficacy by increasing the speed of the search exponentially in the case of an uncharted system. The search method in photosynthesis, which begins with the pigment antenna and progresses to the reaction center in some publications, is also paralleled with the Grover form search. However, the algorithm of subdivision 3.2 is more widely used to describe the high efficacy of the energy transfer process in photosynthesis (Figure 4.11) (Johnson, 2009; Kathrada, 2009).

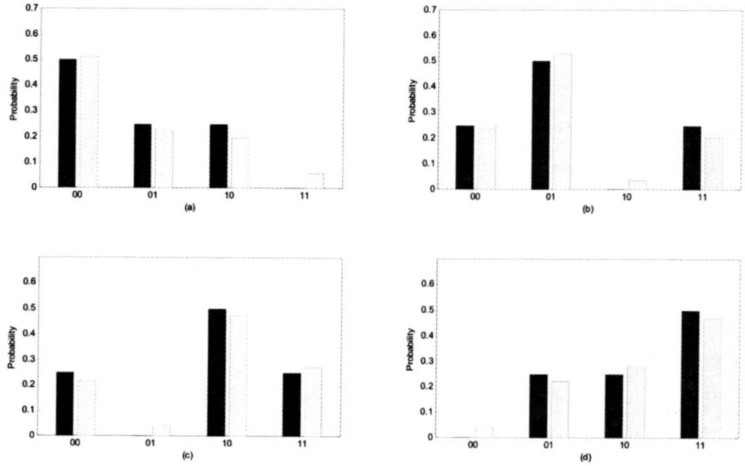

Figure 4.11. Experimental outcomes of the SKW algorithm. (a), (b), (c), (d) relate to the cases of finding |00›12, |01›12, |10 ›12 and |11›12. The theoretical prediction is shown by the blue (dark) bars, whereas the experimental analog is represented by the gray (light) bars. Quantum walk and photosynthesis.

Source: *https://www.intechopen.com/books/search-algorithms-and-applications/search-via-quantum-walk.*

In photosynthesis, energy is collected by dyes and delivered to the reaction center, where it is turned into an electron transfer process, and the chemical energy is initiated. Photosynthesis is a natural process that occurs in nature. The following figure 4.11 depicts a model of energy transmission from the transmitters to the reaction center. Model of energy transmission from the transmitters to the reaction center. In the past, it was widely believed that the excitons emitted by the antennas made their way to the reaction center like the classical random walk. However, this is being challenged by a new conceptual model and recent tests, which show that quantum consistency is related to the energy transfer mechanism, which helps to explain photosynthesis's high effectiveness (Watts, 1999, 2000).

Another remarkable finding is the contribution of the surrounding ecosystem to the quantum walk transmission in photosynthetic energy transfer, which was previously unknown. If it is determined that the environment is the primary cause of decoherence in quantum systems, and that decoherence is the primary impediment to the development of a quantum computer that outperforms the conventional computer, The interplay between the unconstrained Schrödinger of the protein complex and the temperature variations in the environment, on the other hand, results in a rise in the energy transfer effectiveness from around 70% to 95% (Figures 4.12 and 4.13) (Watts et al., 1998).

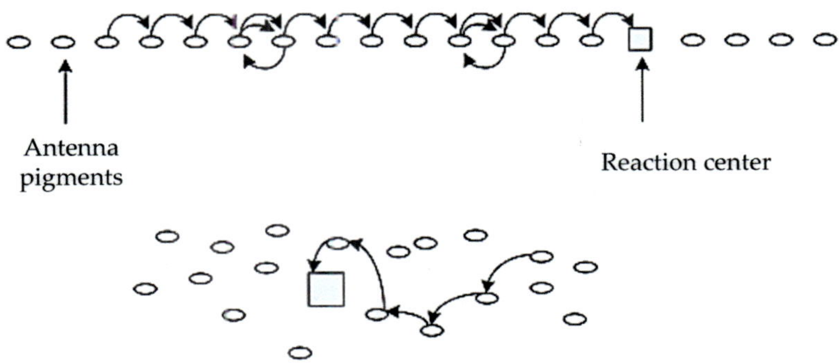

Figure 4.12. Models for the arrangement of antennas. The antennas are represented by the circle, while the rectangle represents the response center. The one-dimensional array model is depicted on the top schematic, while the three-dimensional array model is depicted on the bottom. Of course, the three-dimensional model is more accurate in representing the actual situation (Blankenship, 2002).

Source: https://www.intechopen.com/books/search-algorithms-and-applications/search-via-quantum-walk.

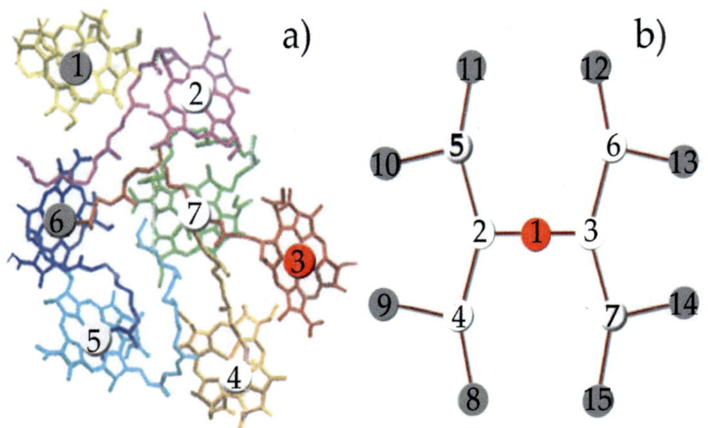

Figure 4.13. (a) Chlorophyll molecules are a kind of phytochrome. It is frequently explored for its simple structure compared to the chlorophyll molecules found in higher plants and algae. The Fenna-Matthews-Olson (FMO) protein

complex is one example. (b) artificial systems characterized by a Hamiltonian with a high degree of tightness.

Source: https://www.intechopen.com/books/search-algorithms-and-applications/search-via-quantum-walk.

Note: Here is an example of a binary tree with four generations. In the background of quantum walk algorithms, it has been hypothesized that some target sites (red) in these structures can be reached at exponentially faster rates than other target sites (Rebentrost, 2009).

4.6. BIOMIMETIC APPLICATION IN SOLAR ENERGY

Because fossil fuels are depleting, energy is becoming a crucial challenge for humans. The direct consumption of solar energy is a widely acknowledged alternative energy consumption strategy since solar energy may be obtained continually. However, at this time, the efficiency with which solar energy can be harvested is still relatively low. Instead, the efficiency of energy transmission in photosynthesis is extremely high: > 90%, and at times even up to 99%, according to some estimates. Suppose the energy transfer efficacy of the solar cell is the same as that of photosynthesis. In that case, the transformation efficiency of the solar cell will grow twice as much as that of photosynthesis does. Understanding the notion of energy transference in photosynthesis will provide a fascinating perspective on the future of energy consumption, and we should strive to achieve this. Many organizations have pledged to develop an artificial photosynthesis system in the past, but no effective enough findings have been acquired thus far. Even though many of them claim to have solved the problem of energy on the planet, scientists are unable to identify when we will be able to get hydrogen to emerge from the water in an expanded manner under the light of the sun, according to their findings (Wasserman and Faust, 1994).

Because several eyes seek the benefits of solar energy, it is not essential to rank them in order of importance. The only thing left for us to do is to bring the notion of photosynthesis to life to use it to our advantage, which we shall accomplish.

REFERENCES

1. Aharonov, Y., Davidovich, L., & Zagury, N., (1993). Quantum random walks. *Phys. Rev. A,* 48, 1687.
2. Ambainis, A., (2003). Quantum walks and their algorithmic applications. *International Journal of Quantum Information, 1*, 507–518.
3. Ambainis, A., (2007). Quantum walk algorithm for element distinctness. *SIAM Journal on Computing, 37*(1), 210–239.
4. Ambainis, A., Kempe, J., & Rivosh, A., (2005). Coins make quantum walks faster. *Proceedings of the 16th Annual ACM-SIAM Symposium on Discrete Algorithms* (pp. 1099–1108). 0-89871-585-7, Vancouver, British Columbia, 2005, Society for Industrial and Applied Mathematics. Philadelphia.
5. Blankenship, R. E., (2002). *Molecular Mechanisms of Photosynthesis*. Wiley-Blackwell, 978-0632043217, Oxford/ Malden.
6. Broome, M. A., Fedrzzi, A., Lanyon, B. P., Kassal, I., Apspuru-Guzik, A., & White, A. G., (2010). Discrete single-photon quantum walks with tunable decoherence. *Phys. Rev. Lett., 104*, 153602.
7. Chandrashekar, C., Srikanth, R., & Laflamme, R., (2008). Optimizing the discrete time quantum walk using a SU(2) coin. *Phys. Rev. A, 77*, 032326.
8. Childs, A. M., & Goldstone, J., (2004). Spatial search by quantum walk. *Phys. Rev. A, 70*, 022314.
9. Childs, A. M., Cleve, R., Deotto, E., Farhi, E., Gutmann, S., & Speilman, D. A., (2003). Exponential algorithmic speedup by quantum walk. *Proceedings of the Thirty-Fifth Annual ACM Symposium on Theory of Computing* (pp. 59–68). 1-58113-674-9, San Diego, CA, USA, ACM, New York.
10. Douglas, B. L., & Wang, J. B., (2009). Efficient quantum circuit implementation of quantum walks. *Phys. Rev. A, 79*, 052335.
11. Du, J. F., Li, H., Xu, X. D., Shi, M. J., Wu, J. H., Zhou, X. Y., & Han, R. D., (2003). Experimental implementation of the quantum random-walk algorithm, *Phys. Rev. A, 67*, 042316.
12. Engel, G. S., Calhoun, T. R., Read, E. L., Ahn, T. K., Mancˇal, T., Cheng, Y. C., Blankenship, R. E., & Fleming, G. R., (2007). Evidence for wavelike energy transfer through quantum coherence in photosynthetic systems. *Nature, 446*, 782.

13. Farhi, E., & Gutmann, S., (1998). Quantum computation and decision trees. *Phys. Rev. A, 58*, 915–928.
14. Grover, L. K., (1997). Quantum mechanics helps in searching for a needle in a haystack. *Phys. Rev. Lett., 79*, 325–328.
15. Hilley, M., Reitzner, D., & Bužek, V., (2009). *Searching Via Walking: How to Find a Marked Subgraph of a Graph Using Quantum Walks*. arXiv, arXiv:0911.1102v1.
16. Johnson, T., (2009). *Simulations of Multi-Waypoint Flocking Problem with Delays and Saturation*. (pp. 1-7)
17. Karski, M., Förster, L., Choi, J. M., Steffen, A., Alt, W., & Meschede, D., (2009). Quantum walk in position space with single optically trapped atoms. *Science, 325*, 174–177.
18. Kathrada, M., (2009). The flexi-PSO: Towards a more flexible particle swarm optimizer. *OPSEARCH, 46*(1), 52–68.
19. Kempe, J., (2003). Quantum random walks - an introductory overview. *Contemporary Physics, 44*(4), 307–327.
20. Kendon, V. M., (2006). A random walk approach to quantum algorithms. *Phil. Trans. R. Soc. A, 364*, 3407–3422.
21. Kendon, V., & Maloyer, O., (2007). *Optimal Computation with Noisy Quantum Walks*. Quantum Information School of Physics & Astronomy University of Leeds Leeds LS2 9JT.
22. Kennedy, J., & Mendes, R., (2002). Population structure and particle swarm performance. In: *Evolutionary Computation, 2002, CEC'02; Proceedings of the 2002 Congress* (Vol. 2, pp. 1671–1676). IEEE.
23. Kennedy, J., & Mendes, R., (2006). Neighborhood topologies in fully informed and best-of-neighborhood particle swarms. *IEEE Transactions on Systems, Man, and Cybernetics, Part C (Applications and Reviews), 36*(4), 515–519.
24. Kennedy, J., (1999). Small worlds and mega-minds: Effects of neighborhood topology on particle swarm performance. In: *Evolutionary Computation, 1999, CEC 99; Proceedings of the 1999 Congress* (Vol. 3, pp. 1931–1938). IEEE.
25. Kennedy, J., (2003). Bare bones particle swarms. In: *Swarm Intelligence Symposium, 2003, SIS'03; Proceedings of the 2003 IEEE* (pp. 80–87). IEEE.
26. Kennedy, J., (2011). Particle swarm optimization. In: *Encyclopedia of Machine Learning* (pp. 760–766). Springer US.

27. Kerdels, J., & Peters, G., (2012). *A Generalized Computational Model for Modeling and Simulation of Complex Systems*. Fernuniversität Hagen.

28. Khan, T. A., Taj, T. A., Asif, M. K., & Ijaz, I., (2012). Modeling of a standard particle swarm optimization algorithm in MATLAB by different benchmarks. In: *Innovative Computing Technology (INTECH), 2012 Second International Conference* (pp. 271–274). IEEE.

29. Khaneja, N., Reiss, T., Kehlet, C., Schulte-Herbrüggen, T., & Glaser, S., (2005). Optimal control of coupled spin dynamics: Design of NMR pulse sequences by gradient ascent algorithms. *J. Magn. Reson., 172*, 296.

30. Kirk, D. B., & Wen-Mei, W. H., (2016). *Programming Massively Parallel Processors: A Hands-on Approach*. Morgan Kaufmann.

31. Kislitsyn, A. A., Kozlova, A. B., Masherov, E. L., & Orlov, Y. N., (2017). *Numerical Algorithm for Self-Consistent Stationary Level for Multidimensional Non-Stationary Time-Series* (No. 1, pp. 14–124). Препринты Института прикладной математики им. МВ Келдыша РАН.

32. Koenig, S., & Sun, X., (2009). Comparing real-time and incremental heuristic search for real-time situated agents. *Autonomous Agents and Multi-Agent Systems, 18*(3), 313–341.

33. Koenig, S., (2004). A comparison of fast search methods for real-time situated agents. In: *Proceedings of the Third International Joint Conference on Autonomous Agents and Multiagent Systems* (Vol. 2, pp. 864–871). IEEE Computer Society.

34. L'Ecuyer, P., & Simard, R., (2007). TestU01: AC library for empirical testing of random number generators. *ACM Transactions on Mathematical Software (TOMS), 33*(4), 22.

35. Li, J., Wan, D., Chi, Z., & Hu, X., (2007). An efficient fine-grained parallel particle swarm optimization method based on GPU-acceleration. *International Journal of Innovative Computing, Information and Control, 3*(6), 1707–1714.

36. Liang, J. J., & Suganthan, P. N., (2005). Dynamic multi-swarm particle swarm optimizer with local search. In: *Evolutionary Computation, 2005; The 2005 IEEE Congress* (Vol. 1, pp. 522–528). IEEE.

37. Liang, J. J., Qin, A. K., Suganthan, P. N., & Baskar, S., (2006). Comprehensive learning particle swarm optimizer for global

optimization of multimodal functions. *IEEE Transactions on Evolutionary Computation, 10*(3), 281–295.

38. Lindholm, E., Nickolls, J., Oberman, S., & Montrym, J., (2008). NVIDIA tesla: A unified graphics and computing architecture. *IEEE Micro, 28*(2).
39. Lu, D., Zhu, J., Zou, P., Peng, X., Yu, Y., Zhang, S., Chen, Q., & Du, J., (2010). Experimental implementation of a quantum random-walk search algorithm using strongly dipolar coupled spins. *Phys. Rev. A, 81*,022308.
40. Luštrek, M., & Bulitko, V., (2006). Lookahead pathology in real-time path-finding. In: *Proceedings of the National Conference on Artificial Intelligence (AAAI), Workshop on Learning for Search* (pp. 108–114).
41. Luštrek, M., (2005). Pathology in single-agent search. In: *Proceedings of Information Society Conference* (pp. 345–348).
42. Mohseni, M., Rebentrost, P., Lloyd, S., & Aspuru-Guzik, A., (2008). Environment-assisted quantum walks in photosynthetic energy transfer. *J. Chem. Phys., 129*, 174106.
43. Nielsen, M. A., & Chuang, I., (2002). *Quantum Computation and Quantum Information.*(Vol.1, pp.1-9)
44. Panitchayangkoon, G., Hayes, D., Fransted, K. A., Caram, J. R., Harel, E., Wen, J. Z., Blankenship, R. W., & Engel, S., (2010). Long-lived quantum coherence in photosynthetic complexes at physiological temperature. *Proc. Natl. Acad. Sci. USA, 107*, 12766–12770.
45. Perets, H. B., Lahini, Y., Pozzi, F., Sorel, M., Morandotti, R., & Silberberg, Y., (2008). Realization of quantum walks with negligible decoherence in waveguide lattices. *Phys. Rev. Lett., 100*, 170506.
46. Pérez, D. C. A., (2007). *Quantum Cellular Automata: Theory and Applications* (p. 61). A Thesis for the Degree of Doctor of Philosophy in Computer Science.
47. Potocek, V., Gábris, A., Kiss, T., & Jex, I., (2009). Optimized quantum random-walk search algorithms on the hypercube. *Phys. Rev. A, 79*, 012325.
48. Rebentrost, P., Mohseni, M., Kassal, I., Lloyd, S., & Aspuru-Guzik, A., (2009). Environment-assisted quantum transport. *New Journal of Physics, 11*, 033003.
49. Reichl, L. E., (1998). *A Modem Course in Statistical Physics*. John Wiley & Sons, Inc., 978-0471595205, New York.

50. Reitzner, D., Hillery, M., Feldman, E., & Bužek, V., (2009). Quantum searches on highly symmetric graphs. *Phys. Rev. A, 79*, 012323.

51. Roberge, V., & Tarbouchi, M., (2012). Efficient parallel particle swarm optimizers on GPU for real-time harmonic minimization in multilevel inverters. In: *IECON 2012–38th Annual Conference on IEEE Industrial Electronics Society* (pp. 2275–2282). IEEE.

52. Rost, R. J., Licea-Kane, B., Ginsburg, D., Kessenich, J., Lichtenbelt, B., Malan, H., & Weiblen, M., (2009). *OpenGL Shading Language*. Pearson Education.

53. Ryan, C., Laforest, M., & Laflamme, R., (2005). Experimental implementation of a discrete-time quantum random walk on an NMR quantum-information processor. *Phys. Rev. A, 72*, 012328.

54. Ryan, C., Negrevergne, C., Laforest, M., Knill, E., & Laflamme, R., (2008). Liquid-state nuclear magnetic resonance as a testbed for developing quantum control methods. *Phys. Rev. A, 78*, 012328.

55. Ryoo, S., Rodrigues, C. I., Baghsorkhi, S. S., Stone, S. S., Kirk, D. B., & Hwu, W. M. W., (2008). Optimization principles and application performance evaluation of a multithreaded GPU using CUDA. In: *Proceedings of the 13th ACM SIGPLAN Symposium on Principles and Practice of Parallel Programming* (pp. 73–82, 1652–1659). ACM.

56. Sadikov, A., & Bratko, I., (2006). Pessimistic heuristics beat optimistic ones in real-time search. *Frontiers in Artificial Intelligence and Applications, 141*, 148.

57. Sanders, J., & Kandrot, E., (2010). *CUDA by Example: An Introduction to General-Purpose GPU Programming*. Addison-Wesley Professional.

58. Schmitz, H., Matjeschk, R., Schneider, C., Gluechert, J., Enderlein, M., Huber, T., & Schaetz, T., (2009). Quantum walk of a trapped ion in phase space. *Phys. Rev. Lett., 103*, 090504.

59. Schreiber, A., Cassemiro, K. N., Potocˇek, V., Gábris, A., Mosley, P. J., Andersson, E., Jex, I., & Silberhorn, C., (2010). Photons walking the line: A quantum walk with adjustable coin operations. *Phys. Rev. Lett., 104*, 050502.

60. Secrest, B. R., & Lamont, G. B., (2003). Visualizing particle swarm optimization-Gaussian particle swarm optimization. In: *Swarm Intelligence Symposium, 2003, SIS'03; Proceedings of the 2003 IEEE* (pp. 198–204). IEEE.

61. Shenvi, N., Kempe, J., & Whaley, K., (2003). Quantum random-walk search algorithm. *Phys. Rev. A, 67*, 052307.
62. Shi, Y., & Eberhart, R. C., (1999). Empirical study of particle swarm optimization. In: *Evolutionary computation, 1999, CEC 99; Proceedings of the 1999 Congress* (Vol. 3, pp. 1945–1950). IEEE.
63. Shi, Y., & Eberhart, R., (1998). A modified particle swarm optimizer. In: *Evolutionary Computation Proceedings, 1998; IEEE World Congress on Computational Intelligence, The 1998 IEEE International Conference* (pp. 69–73). IEEE.
64. Shue, L. Y., & Zamani, R., (1993). An admissible heuristic search algorithm. In: *International Symposium on Methodologies for Intelligent Systems* (pp. 69–75). Springer, Berlin, Heidelberg.
65. Shue, L. Y., Li, S. T., & Zamani, R., (2001). An intelligent heuristic algorithm for project scheduling problems. In: *Proceedings of the 32nd Annual Meeting of the Decision Sciences Institute*.
66. Simard, R., & L'Ecuyer, P., (2011). Computing the two-sided Kolmogorov-Smirnov distribution. *Journal of Statistical Software, 39*(11), 1–18.
67. Suryaprakash, N., (2000). Liquid crystals as solvents in NMR: Spectroscopy current developments in structure determination. *Current Organic Chemistry, 4*, 85–103.
68. Tulsi, A., (2008). Faster quantum-walk algorithm for the two-dimensional spatial search, *Phys. Rev. A, 78*, 012310.
69. Wasserman, S., & Faust, K., (1994). *Social Network Analysis: Methods and Applications* (Vol. 8). Cambridge University Press.
70. Watts, D. J., & Strogatz, S. H., (1998). Collective dynamics of 'small-world' networks. *Nature, 393*(6684), 440.
71. Watts, D. J., (1999). *Small Worlds: The Dynamics of Networks Between Order and Randomness*. Princeton University Press.
72. Watts, D. J., (2000). *Small Worlds: The Dynamics of Networks Between Order and Randomness*. Princeton: Princeton University Press.
73. Zähringer, F., Kirchmair, G., Gerritsma, R., Solano, E., Blatt, R., & Roos, C. F., (2010). Realization of a quantum walk with one and two trapped ions. *Phys. Rev. Lett., 104*, 100503.

CHAPTER 5

AN INTRODUCTION TO HEURISTIC ALGORITHMS

CONTENTS

5.1. INTRODUCTION

Computers are currently being used to address problems that are extremely complicated. It is necessary to build an algorithm, however, in order to deal with an issue effectively. On rare occasions, the human brain is unable to do this task successfully. Furthermore, accurate algorithms may take hundreds of years or even decades to deal with daunting difficulties. Consequently, heuristic algorithms that provide approximate solutions yet have tolerable time and space complexity play an essential role in achieving a satisfactory result (Lin and Kernighan, 1971). The essential underlying notions and the ranges of application of heuristics are discussed in detail in this chapter on heuristics. A similar approach was used by us in which we presented in further detail certain novel heuristic techniques, specifically support vector machines (SVMs) and evolutionary algorithms (EAs) (Figure 5.1) (Lin and Kernighan, 1973).

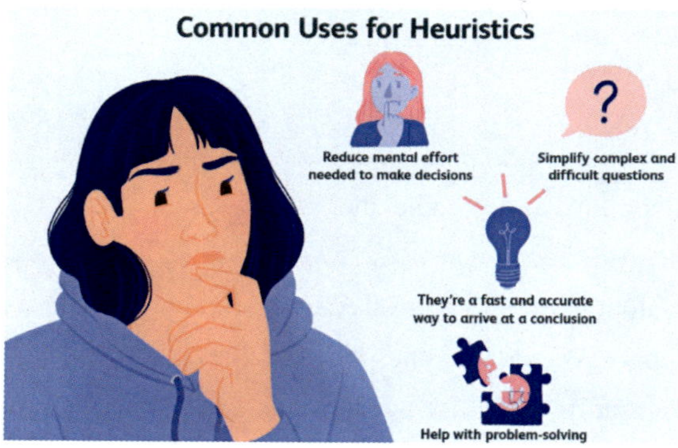

Figure 5.1. Common uses of heuristics.

Source: https://www.verywellmind.com/what-is-a-heuristic-2795235.

Complexity estimation, algorithm validation, and optimization are the most important subjects in a wide range of computation-related topics. These problems are addressed in large part of theoretical computer science. In general, job complexity is evaluated by examining the most appropriate computational resources, such as execution space and time. Extending problems that can be solved in a limited amount of space and time into well-defined categories is a difficult undertaking, but it can significantly reduce the amount of time and money spent on algorithm development. Algorithm development was the subject of a large number of studies. In the literature, there is a brief historical summary of the core concerns in computing theory.

We don't have a clear concept of algorithm and complexity (Cormen et al., 1989).

Modern challenges are extremely complicated and involve the analysis of massive data volumes. Even if an accurate method can be developed, its time or space complexity may be unacceptably high. However, finding a partial or approximate answer is frequently sufficient. This type of admittance entails a set of procedures for dealing with the issue. We discuss heuristic algorithms, which offer approximations for solving optimization issues. The goal of these issues is to discover the best answer out of all feasible solutions, which is explicitly the one that maximizes or reduces an objective function. The objective function is a function that is used to evaluate the quality of a created solution. Many real-world problems can be described as optimization problems. A search space is a collection of all possible solutions to a problem, and optimization techniques are frequently referred to as search algorithms.

Approximate algorithms raise the intriguing question of evaluating the quality of the solutions they provide. Given that the ideal solution is frequently unknown, this problem could pose a significant challenge in terms of strong mathematical analysis. In terms of quality, the heuristic algorithm's goal is to find as excellent a solution as feasible for all occurrences of the problem. There are a variety of generic heuristic approaches that can be used to solve a variety of problems (Figure 5.2) (Gillett and Miller, 1974; Nawaz et al., 1983).

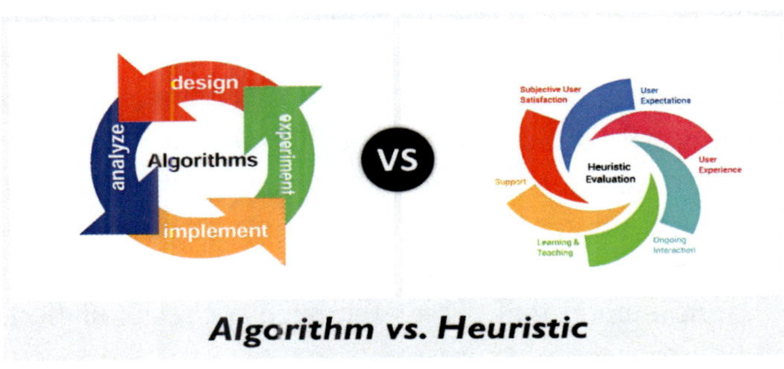

Figure 5.2. Comparison between conventional algorithms and heuristic algorithms.

Source: https://www.differencebtw.com/difference-between-algorithm-and-heuristic.

5.2. ALGORITHMS AND COMPLEXITY

It is difficult to comprehend the vast array of current computational jobs, let alone the vast array of algorithms devised to tackle them. Heuristic algorithms are algorithms that either has a solution or supply a solution that is close to the correct answer, but not for all circumstances of the problem. This group includes a diverse range of strategies that are based on both traditional and specialized procedures, as well as hybrid approaches. To begin, we will summarize the fundamental concepts of classical search algorithms (Lee and Geem, 2005). The most basic of search algorithms are referred to as 'exhaustive search.' It runs through all of the possible solutions from a pre-programed set and then selects the best of them.

Local search is a type of exhaustive search that just focuses on a small portion of the search space. It is a subset of exhaustive search. Different approaches might be taken to systematizing local search. This category includes the most popular hill-climbing techniques. Such algorithms constantly replace the current solution with the best of its neighbors if the best of its neighbors is preferable to the existing solution. Consider, for example, the use of the hill-climbing approach to solving the problem of intragroup imitation in a multimedia distribution service based on a peer-to-peer network.

Generally speaking, divide and conquer algorithms tend to break down a large problem into smaller problems that are easier to tackle. The solutions to the minor difficulties should be analogous to the answer to the major problem. Even though this technique appears to be promising, its application is limited because there is a limited number of problems that can be efficiently partitioned and combined in this manner. The branch-and-bound approach is a frantic enumeration of the search space that results in an error. It enumerates but has a strong tendency to exclude regions of the search space that do not contain the optimal answer on a consistent basis. By retaining the solutions to sub-problems, dynamic programming provides an efficient search method that avoids the need for recalculation. The fact that the solution approach is expressed as recursion is the most important factor to consider while employing this technique. The 'greedy' technique is a popular way of creating a sequential space of solutions. It is based on the simple premise of making the (local) best decision at each stage of the algorithm in order to find the global best of some objective function (Campbell et al., 1970; Gupta, 1971).

Heuristic algorithms are typically used to address issues that are difficult to solve, such as those involving large amounts of data. The categories of temporal complexity are explained in order to different situations according to their "hardness." From the standpoint of deterministic Turing machines, Category P consists of all problems that can be solved in polynomial time by a deterministic Turing computer, given the magnitude of the input. Computing complexity and algorithms are concepts that are validated through the usage of Turing machines, which are perceptions of how computers work. In the case of non-deterministic Turing machines, category NP includes all issues for which a solution may be found in polynomial time on a Turing machine. Because such a machine does not exist, it is reasonable to claim that an exponential algorithm may be inscribed for an NP problem; nevertheless, there is no assurance that a polynomial algorithm will occur or that it will not occur. There are issues in the category NP-complete, which is a subclass of NP. Some of them, for example, could be transformed into polynomial algorithms that could be used to tackle all other NP issues. Finally, the category NP-hard can be taken to be the category of problems that are either NP-complete or significantly more difficult. NP-hard problems have many characteristics with NP-complete problems, but they do not necessarily belong to class NP; for example, class NP-hard problems contain problems for which no algorithms can be provided at all, as well as problems for which no algorithms can be provided at all (Armour and Buffa, 1963; Christofides, 1976).

Before applying a heuristic algorithm, we must first determine whether the problem is NP-complete or NP-hard in order to ensure that the method is appropriate for the situation. It is highly likely that there are no polynomial algorithms for tackling such issues; as a result, heuristics are created for situations with sufficiently large inputs (Jaw et al., 1986; Mahdavi et al., 2007; Omran and Mahdavi, 2008).

5.3. HEURISTIC TECHNIQUES

Although dynamic programming and branch-and-bound techniques are successful, their time complexity is frequently excessively high and unbearable for NP-complete jobs. The hill-climbing algorithm is effective, but it suffers from early convergence, which is a big drawback. It always finds the nearest low-quality local optima because it is "greedy." Modern heuristics are designed to reduce or eliminate this disadvantage (Figure 5.3) (Lee and Geem, 2004; Kennedy, 2011).

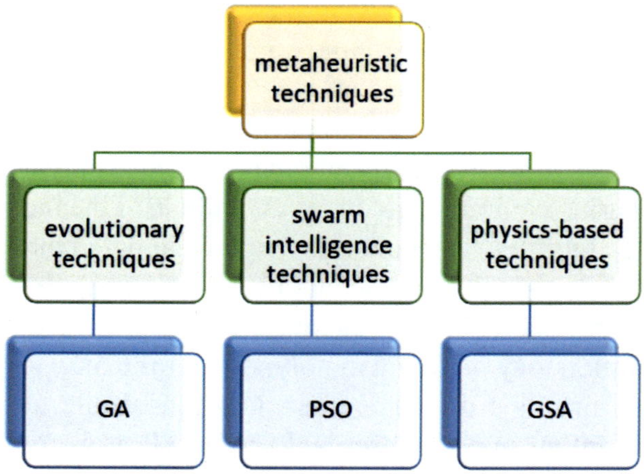

Figure 5.3. Different meta-heuristic techniques.

Source: https://www.researchgate.net/figure/Categories-of-meta-heuristic-techniques_fig3_350874721.

Developed in 1983, the simulated annealing technique makes use of a way comparable to hill-climbing, but it only accepts solutions that are marginally better than the present answer. The likelihood of such acceptance diminishes with the passage of time.

It is the concept of evading local optima via expanding memory structures that are encompassed by the tabu search algorithm. The difficulty with simulated annealing is that after a "jump," the algorithm can just repeat its own route, which is undesirable. The use of tabu search eliminates the repeating of changes that have recently been made.

In 1989, the concept of swarm intelligence was introduced (Eberhart et al., 2001). An example of this is the study of coupled behavior in self-organized, decentralized systems, which is the basis for the approach of simulated intelligence. There are several types of optimization techniques available, but two of the most successful are particle swarm optimization (PSO) and ant colony optimization (ACO) (ACO). Artificial ants generate solutions in ACO by roaming about on the problem chart and modifying it in such a way that future ants can construct even better solutions in the future. Problems in which the best answer can be represented as a point or as a surface in n-dimensional space are dealt with by PSO algorithms. Among the many advantages of swarm intelligence approaches is the fact that they are extraordinarily immune to the problem of local optimality. EAs are successful in their attempts at premature convergence because they perceive a large number of solutions at the same time (Kirkpatrick et al., 1983; Geem, 2006). Later on, we shall go into greater detail about this particular category of algorithms.

Biological neuron systems are the source of inspiration for neural networks. They are made up of individual units known as neurons, as well as connections between them. Following extensive preparation on a hypothetical dataset, it is possible for neural networks to generate predictions about cases that are not included in the preparation set. In actuality, Neural Networks do not perform consistently effectively since they suffer from substantial difficulties of overfitting and underfitting on a regular basis. These issues are related to the precision with which predictions are made. If a system is not complex enough, it may be necessary to reduce the rules that the data must follow. On the other hand, if a system is overly complex, it may take into account the noise that is typically present in the preparation data set while inferring the rules. In both cases, the value of a prediction after preparation has been diminished by the preparation. Additionally, the problem of early convergence is significant for Neural Networks (Holland, 1992; Helsgaun, 2000; Geem et al., 2005).

SVMs are a type of neural network that incorporates the concepts of reinforcement learning and neural networks. They are successful in avoiding early convergence because they take into consideration the convex goal function; as a result, only one optimum exists. The traditional divide and conquer strategy give an elegant solution for situations that are easily distinguished. It becomes an immensely strong tool when used in conjunction with SVMs, which provide effective classification. Later on, we will discuss SVM classification trees and which applications are now providing a potential object for further research and development (Johnson and McGeoch, 1997).

In the literature, there is a comparative analysis and explanation of simulated annealing, neural networks, EAs, and tabu search, among other techniques (Reinelt, 1991; Pham and Karaboga, 2012).

5.4. EVOLUTIONARY ALGORITHMS (EAS)

Evolutionary algorithms (EAs) are methods for investigating the solution to an optimization problem that use biological evolution concepts such as recombination, reproduction, and mutation. They use the survival principle to generate steady estimates to the optimum from a set of feasible solutions.

The process of selecting people in accordance with their objective function, known as fitness for EAs, also breeds them together with operators inspired by genetic processes, resulting in a distinct set of approximations (Taillard, 1990, 1993). This process results in the evolution of a population of people who are better suited to their environment than their forefathers (Figure 5.4).

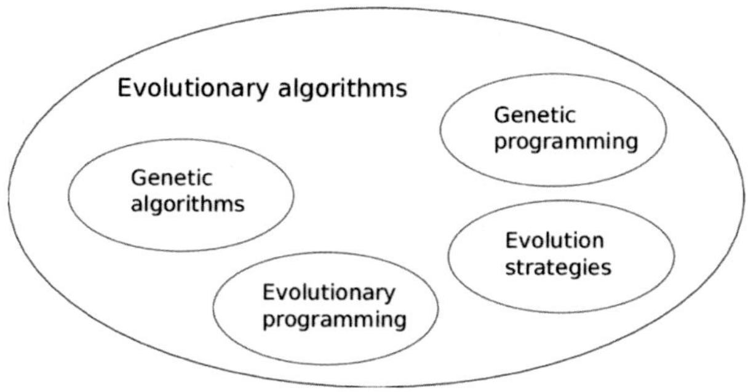

Figure 5.4. Branches of evolutionary algorithms.

Source: https://www.researchgate.net/figure/The-classification-of-evolutionary-algorithms_fig1_324994158.

The main circle of EAs contains the following steps:

- Initialize and assess the initial population;
- Execute competitive selection;
- Apply genetic operators to produce new solutions;
- Calculate solutions in the population;
- Start over from point 2 and reprise until some convergence standard is satisfied.

Although evolutionary approaches have a basic concept, the specifics of their execution and the situations in which they are employed can differ. In the form of computer programs, genetic programming looks for solutions. The ability to solve a computational problem is what determines their fitness. The difference between evolutionary and genetic programming is that the former secures the program's organization while allowing numerical parameters to alter. The self-adaptive mutation rates are taken into account in the evolution approach, which uses vectors of real numbers as solutions (Palmer, 1965; Dannenbring, 1977).

Among EAs, genetic algorithms (GAs) have the highest success. John Holland (1992) examined these methods and found them to be extremely useful. GAs are founded on the idea that the role of mutation in the development of a person infrequently occurs. Therefore, they rely on recombination operators most of the time. They try to solve problems by using a series of numbers, usually binary, to solve the problems (Reeves, 1995; Ruiz and Maroto, 2005).

The most common application of GAs is to solve optimization problems requiring a large number of high-performance computing resources. Wu et al. (2005) for example, looked into the question of effective resource allocation in the context of the PCMA system. Its goal is to make packet-switching wireless cellular networks as capable as possible. The most important consideration when allocating resources is to reduce the amount of bandwidth that must be provided. It has been determined that the problem falls under the NP-hard category. Instead of using a greedy search, the authors utilized a genetic algorithm. The ideal cellular system, consisting of a single base station capable of supporting m connections at one UB per second and arranged in a cluster of B cells, has been computer-imitated. As a result, a genetic algorithm can be used to optimize system capacity utilization (Clarke and Wright, 1964; Gendreau et al., 1994).

Because of the significant increase in consumer demand for mobile phones and computers, network optimization issues have become extremely important. EAs are often employed in this area since they are a standard technique that can be easily adapted under varied conditions (Wren and Holliday, 1972; Osman, 1993).

Consider the adaptive mesh problem (AMP), which seeks to reduce the number of cellular network base stations required to cover a given area. AMP is NP-hard, just like the prior discussed issue. One of the evolutionary strategies used to tackle this challenge is HIES (Gaskell, 1967; Creput et al., 2005). It refers to an evolutionary algorithm that borrows properties from fine-grained or cellular GAs and island model GAs. The original problem has been transformed into a geometric interconnecting generation problem, with specialized genetic operators such as crossover, macro mutation, and micro mutation changing a variety of hexagonal cells. Regular honeycomb was initially transformed to irregular mesh, which better reflects real-world conditions (Mole and Jameson, 1976; Mahdavi et al., 2007). The goal of reducing the overall number of base stations has been accomplished. In work devoted to machine learning theory difficulties (Divina and Marchiori, 2005), some further examples of EAs were discovered.

5.5. SUPPORT VECTOR MACHINES (SVMS)

The challenge of selecting preferred functions based on empirical data is addressed by statistical learning theory. Its core issue is generalization, which entails inferring rules for relevant features from a small set of data. SVMs is the most widely used technology in this sector nowadays (Figure 5.5) (Geem et al., 2002; Fesanghary, 2008; Geem, 2008).

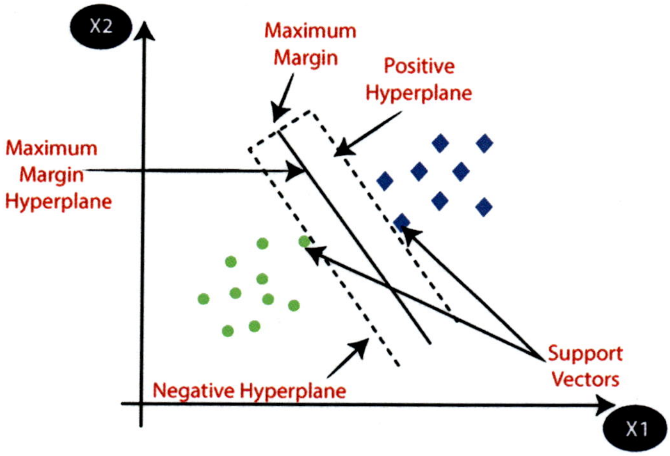

Figure 5.5. An example of support vector machine.

Source: https://www.javatpoint.com/machine-learning-support-vector-machine-algorithm.

Vapnik established the fundamental ideas of SVMs and their applications (2013). A wide range of applications arose almost immediately as a result of their enticing potential. SVMs diverged from the principle of empirical risk minimization (ERM), which is exemplified by conservative neural networks, and took into account the principle of structural risk minimization (SRM), which diminishes an upper bound on the anticipated risk, rather than the principle of ERM (Gunn, 1998; Kim et al., 2001). SVMs are useful for tasks such as regression and classification because they are based on the concept of the optimal separator. The classification problem can be expressed as a problem of dividing a dataset into groups using the functions that are invoked by the instances that are now available. Classifiers will be used to refer to these types of functions (Nawaz et al., 1983; Widmer and Hertz, 1989; Hoogeveen, 1991).

During the course of a regression task, we must determine whether the dependent variable (y) is functionally dependent on a group of independent factors (x). On the basis of this assumption, a deterministic function (f) denoting the related to the dependent and independent variables is used to model the relationship between the two variables.

Contemplate the problem of splitting the set of vectors taken from two classes:

$\{(x^1,y^1), \ldots, (x^l,y^l)\}, x \in R^n, y \in \{-1,1\}$

with a hyperplane,

hw, xi + b = 0

where; w and b are parameters; hw, xi denotes inner product.

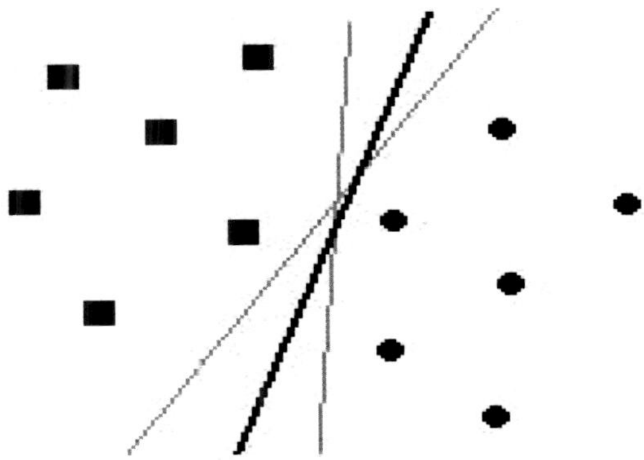

Figure 5.6. Illustration of the classification problem.

Source: https://www.researchgate.net/publication/228573156_An_introduction_to_heuristic_algorithms.

The goal is to separate categories without flaws using the hyperplane while maximizing the space here between the closest vector and the hyperplane. The term "optimal separating hyperplane" refers to such a hyperplane. The best splitting hyperplane, according to Gunn's findings (1998), minimizes.

$$\Phi(w) = \frac{1}{2}\|w\|^2 \qquad (1)$$

Under the constraints:

y^i[hw,xi + b] $\geq$ 1 (2)

The saddle notion of the Lagrangian functional is used to specify the solution to the optimization problems (1) and (2), respectively. Points with non-zero Lagrange multipliers are referred to as Support Vectors, and they are utilized to define the classifier that is formed from the data. Most of the time, the support vectors data set is only a modest part of the preparation data set. This fact contributes to the most attractive aspect of SVMs, which is their low complexity (Ignall and Schrage, 1965; Lee, 1967).

When preparation data is not linearly divided, there are two options: the first is to introduce an additional function concurrently with misclassification, and the second is to practice a more sophisticated function to determine the border of the preparation data set. Furthermore, the optimization problem at large is posed in order to reduce the classification error in furthermore to the bound on the classification classifier's VC dimension (see below). VC dimension of a group of functions is (p) if and only if there exists a group of (p) points, such that these points can be divided in all 2p possible conformations using these functions, and there is no group of (q) points, such that these points cannot be divided in all 2p possible conformations using these functions, and this property is maintained (Buffa, 1964; Nugent et al., 1968).

In this essay, it is not possible to cover the entire large and intricate notion of SVMs. In recent years, a plethora of methods based on the concepts of SVMs has been developed and refined. Among these algorithms, the SVM classification tree algorithm has been effectively employed in image and text classification (Armour, 1974; Kusiak and Heragu, 1987).

In a classification tree, there are internal and external nodes that are linked together by branches. Specifically, each internal node performs a split function, which divides a preparatory data set into two disjoint subgroups, and each external node contains a label specifying which class the particular feature vector should be classified into (Tate and Smith, 1995; Meller and Gau, 1996). It was recently discovered that this technique could be used to deal with the classification intricacy of the membership verification problem, which is a common problem in digital security systems. Specifically, it seeks to distinguish between the membership class (M) and the non-membership class (GM) inside the human group (G). The competence of an SVM classifier is achieved by the use of two partitioned subdivisions, and the competent SVM tree is then applied to the task of determining the membership of an unidentified person in a group. The experimental results have demonstrated that the proposed strategy exhibits greater robustness and performance than earlier techniques compared to the literature (Hundal and Rajgopal, 1988; Ho and Chang, 1991).

5.6. CURRENT TRENDS

Heuristics are approximation methods for solving optimization issues, as described in this chapter. Heuristic algorithms are typically developed with a low time complexity and used for difficult tasks. Basic modern and classic heuristic strategies were briefly defined. SVMs and EAs were discussed in greater depth. They have garnered a lot of popularity as a result of their outstanding qualities. New research findings support the idea that their applications can be significantly expanded in the future (Arora, 1998; Koulamas, 1998).

This chapter appears to be incomplete. It would be interesting to conduct a more thorough investigation of heuristics and compare the accuracy and implementation complexity of various approximate methods. However, because of the large amount of data, this job will be difficult to complete. We didn't even go over a major topic in heuristic algorithms, planning, and scheduling theory. However, we believe that our work will demonstrate the critical importance of heuristics in modern computer science (Rosenkrantz et al., 2009).

REFERENCES

1. Armour, G. C., & Buffa, E. S., (1963). A heuristic algorithm and simulation approach to relative location of facilities. *Management Science, 9*(2), 294–309.
2. Armour, G. C., (1974). In: Elwoods, B., (ed.), *Operations and Systems Analysis: A Simulation Approach*, 292.
3. Arora, S., (1998). Polynomial time approximation schemes for Euclidean traveling salesman and other geometric problems. *Journal of the ACM (JACM), 45*(5), 753–782.
4. Buffa, E. S., (1964). Allocating facilities with CRAFT. *Harvard Business Review, 42*(2), 136–158.
5. Campbell, H. G., Dudek, R. A., & Smith, M. L., (1970). A heuristic algorithm for the n job, m machine sequencing problem. *Management Science, 16*(10), B-630.
6. Christofides, N., (1976). *Worst-Case Analysis of a New Heuristic for the Travelling Salesman Problem (No. RR-388)*. Carnegie-Mellon Univ Pittsburgh Pa Management Sciences Research Group.
7. Clarke, G., & Wright, J. W., (1964). Scheduling of vehicles from a central depot to a number of delivery points. *Operations Research, 12*(4), 568–581.
8. Cormen, T. H., & Leiserson, C. E., (1989). In: Rivest, R. L., (ed.), *Introduction to Algorithms*. MIT Press. Cambridge, Massachusetts London, England.
9. Creput, J. C., Koukam, A., Lissajoux, T., & Caminada, A., (2005). Automatic mesh generation for mobile network dimensioning using evolutionary approach. *IEEE Transactions on Evolutionary Computation, 9*(1), 18–30.
10. Dannenbring, D. G., (1977). An evaluation of flow shop sequencing heuristics. *Management Science, 23*(11), 1174–1182.
11. Divina, F., & Marchiori, E., (2005). Handling continuous attributes in an evolutionary inductive learner. *IEEE Transactions on Evolutionary Computation, 9*(1), 31–43.
12. Eberhart, R. C., Shi, Y., & Kennedy, J., (2001). *Swarm Intelligence (The Morgan Kaufmann Series in Evolutionary Computation), 1,* 1–20.
13. Fesanghary, M., Mahdavi, M., Minary-Jolandan, M., & Alizadeh, Y., (2008). Hybridizing harmony search algorithm with sequential

quadratic programming for engineering optimization problems. *Computer Methods in Applied Mechanics and Engineering, 197*(33–40), 3080–3091.

14. Gaskell, T. J., (1967). Bases for vehicle fleet scheduling. *Journal of the Operational Research Society, 18*(3), 281–295.
15. Geem, Z. W., (2006). Optimal cost design of water distribution networks using harmony search. *Engineering Optimization, 38*(03), 259–277.
16. Geem, Z. W., (2008). Novel derivative of harmony search algorithm for discrete design variables. *Applied Mathematics and Computation, 199*(1), 223–230.
17. Geem, Z. W., Kim, J. H., & Loganathan, G. V., (2001). A new heuristic optimization algorithm: Harmony search. *Simulation, 76*(2), 60–68.
18. Geem, Z. W., Kim, J. H., & Loganathan, G. V., (2002). Harmony search optimization: Application to pipe network design. *International Journal of Modeling and Simulation, 22*(2), 125–133.
19. Geem, Z. W., Lee, K. S., & Park, Y., (2005). Application of harmony search to vehicle routing. *American Journal of Applied Sciences, 2*(12), 1552–1557.
20. Gendreau, M., Hertz, A., & Laporte, G., (1994). A tabu search heuristic for the vehicle routing problem. *Management Science, 40*(10), 1276–1290.
21. Gillett, B. E., & Miller, L. R., (1974). A heuristic algorithm for the vehicle-dispatch problem. *Operations Research, 22*(2), 340–349.
22. Gunn, S. R., (1998). Support vector machines for classification and regression. *ISIS Technical Report, 14*(1), 5–16.
23. Helsgaun, K., (2000). An effective implementation of the Lin–Kernighan traveling salesman heuristic. *European Journal of Operational Research, 126*(1), 106–130.
24. Ho, J. C., & Chang, Y. L., (1991). A new heuristic for the n-job, M-machine flow-shop problem. *European Journal of Operational Research, 52*(2), 194–202.
25. Holland, J. H., (1992). *Adaptation in Natural and Artificial Systems: An Introductory Analysis with Applications to Biology, Control, and Artificial Intelligence*. MIT Press.
26. Hoogeveen, J. A., (1991). Analysis of Christofides' heuristic: Some paths are more difficult than cycles. *Operations Research Letters, 10*(5), 291–295.

27. Hundal, T. S., & Rajgopal, J., (1988). An extension of palmer's heuristic for the flow shop scheduling problem. *International Journal of Production Research, 26*(6), 1119–1124.
28. Ignall, E., & Schrage, L., (1965). Application of the branch and bound technique to some flow-shop scheduling problems. *Operations Research, 13*(3), 400–412.
29. Jaw, J. J., Odoni, A. R., Psaraftis, H. N., & Wilson, N. H., (1986). A heuristic algorithm for the multi-vehicle advance request dial-a-ride problem with time windows. *Transportation Research Part B: Methodological, 20*(3), 243–257.
30. Johnson, D. S., & McGeoch, L. A., (1997). The traveling salesman problem: A case study in local optimization. *Local Search in Combinatorial Optimization, 1*, 215–310.
31. Kennedy, J., (2011). Particle swarm optimization. In: *Encyclopedia of Machine Learning* (pp. 760–766). Springer US.
32. Kim, J. H., Geem, Z. W., & Kim, E. S., (2001). Parameter estimation of the nonlinear Muskingum model using harmony search. *JAWRA Journal of the American Water Resources Association, 37*(5), 1131–1138.
33. Kirkpatrick, S., Gelatt, C. D., & Vecchi, M. P., (1983). Optimization by simulated annealing. *Science, 220*(4598), 671–680.
34. Koulamas, C., (1998). A new constructive heuristic for the flowshop scheduling problem. *European Journal of Operational Research, 105*(1), 66–71.
35. Kusiak, A., & Heragu, S. S., (1987). The facility layout problem. *European Journal of Operational Research, 29*(3), 229–251.
36. Lee, K. S., & Geem, Z. W., (2004). A new structural optimization method based on the harmony search algorithm. *Computers & Structures, 82*(9, 10), 781–798.
37. Lee, K. S., & Geem, Z. W., (2005). A new meta-heuristic algorithm for continuous engineering optimization: Harmony search theory and practice. *Computer Methods in Applied Mechanics and Engineering, 194*(36–38), 3902–3933.
38. Lee, R. C., (1967). CORELAP-computerized relationship layout planning. *Jour. Ind. Engg., 8*(3), 195–200.

39. Lin, S., & Kernighan, B. W., (1971). A heuristic technique for solving a class of combinatorial optimization problems. In: *Princeton Conference on System Science*.
40. Lin, S., & Kernighan, B. W., (1973). An effective heuristic algorithm for the traveling-salesman problem. *Operations Research, 21*(2), 498–516.
41. Mahdavi, M., Fesanghary, M., & Damangir, E., (2007). An improved harmony search algorithm for solving optimization problems. *Applied Mathematics and Computation, 188*(2), 1567–1579.
42. Meller, R. D., & Gau, K. Y., (1996). The facility layout problem: Recent and emerging trends and perspectives. *Journal of Manufacturing Systems, 15*(5), 351–366.
43. Mole, R. H., & Jameson, S. R., (1976). A sequential route-building algorithm employing a generalized savings criterion. *Journal of the Operational Research Society, 27*(2), 503–511.
44. Nawaz, M., Enscore, Jr. E. E., & Ham, I., (1983). A heuristic algorithm for the m-machine, n-job flow-shop sequencing problem. *Omega, 11*(1), 91–95.
45. Nugent, C. E., Vollmann, T. E., & Ruml, J., (1968). An experimental comparison of techniques for the assignment of facilities to locations. *Operations Research, 16*(1), 150–173.
46. Omran, M. G., & Mahdavi, M., (2008). Global-best harmony search. *Applied Mathematics and Computation, 198*(2), 643–656.
47. Osman, I. H., (1993). Metastrategy simulated annealing and tabu search algorithms for the vehicle routing problem. *Annals of Operations Research, 41*(4), 421–451.
48. Palmer, D. S., (1965). Sequencing jobs through a multi-stage process in the minimum total time—A quick method of obtaining a near optimum. *Journal of the Operational Research Society, 16*(1), 101–107.
49. Pham, D., & Karaboga, D., (2012). *Intelligent Optimization Techniques: Genetic Algorithms, Tabu Search, Simulated Annealing and Neural Networks*. Springer Science & Business Media.
50. Reinelt, G., (1991). TSPLIB—A traveling salesman problem library. *ORSA Journal on Computing, 3*(4), 376–384.
51. Rosenkrantz, D. J., Stearns, R. E., & Lewis, P. M., (2009). An analysis of several heuristics for the traveling salesman problem. In: *Fundamental Problems in Computing* (pp. 45–69). Springer, Dordrecht.

52. Ruiz, R., & Maroto, C., (2005). A comprehensive review and evaluation of permutation flowshop heuristics. *European Journal of Operational Research, 165*(2), 479–494.
53. Tate, D. M., & Smith, A. E., (1995). Unequal-area facility layout by genetic search. *IIE Transactions, 27*(4), 465–472.
54. Vapnik, V., (2013). *The Nature of Statistical Learning Theory*. Springer science & business media.
55. Widmer, M., & Hertz, A., (1989). A new heuristic method for the flow shop sequencing problem. *European Journal of Operational Research, 41*(2), 186–193.
56. Wren, A., & Holliday, A., (1972). Computer scheduling of vehicles from one or more depots to a number of delivery points. *Journal of the Operational Research Society, 23*(3), 333–344.
57. Wu, X., Sharif, B. S., & Hinton, O. R., (2005). An improved resource allocation scheme for plane cover multiple access using genetic algorithm. *IEEE Transactions on Evolutionary Computation, 9*(1), 74–81.

CHAPTER 6

MACHINE LEARNING ALGORITHMS

CONTENTS

6.1. INTRODUCTION

The field of statistics known as computational learning theory studies the computational analysis and performance of machine learning methods. Such algorithms are developed in machine learning to assist the computer in learning. Learning does not always imply consciousness; identifying statistical symmetries or patterns in a set of data is also a form of learning. Human learning algorithms bear no resemblance to machine learning algorithms. Machine learning methods, on the other hand, could provide insight into the relative difficulty of learning in various situations (Figure 6.1).

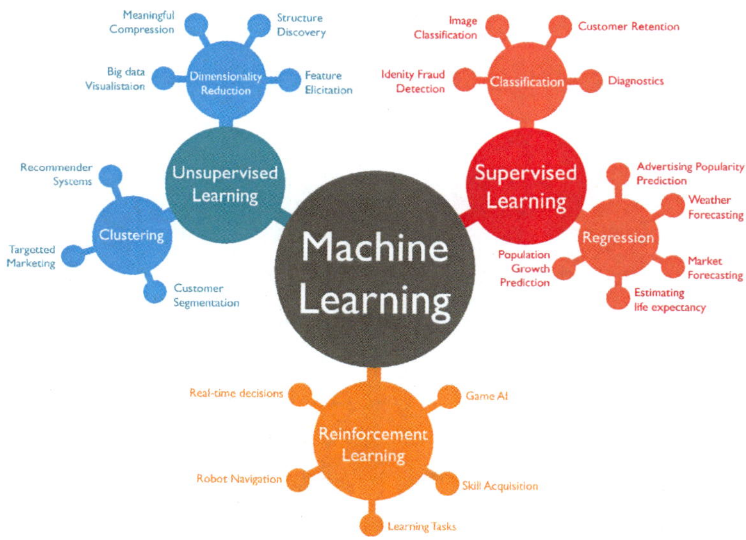

Figure 6.1. Different types of machine learning algorithms.

Source: https://towardsdatascience.com/machine-learning-algorithms-in-lay-mans-terms-part-1-d0368d769a7b?gi=3f432d1ebd11.

Learning algorithms are classified into multiple groups based on the algorithm's preferred outcome. The common learning algorithm types are:

- **Supervised Learning:** The algorithm creates a function that maps the inputs to the expected outputs. One of the classic forms of this learning is the categorization issue. The learner must assimilate information to evaluate the performance of a function in a classification issue. The function displays a vector from many classes by considering various input-output specimens of the function.

- **Unsupervised Learning:** These models a set of inputs. The instances that are labeled are not available.
- **Semi-Supervised Learning:** A suitable function or classifier is produced as a result of combining the labeled and unlabeled instances in this learning process.
- **Reinforcement Learning:** Given the observation of the world, in this case, the algorithm develops a strategy for how to carry out the task. In every action, there is a corresponding effect on the environment, and the environment offers feedback on these effects. The learning algorithm is then escorted by the feedback.
- **Transduction:** This is like supervised learning, except that no function is expressly established. Transduction learning aims to predict new outcomes based on previously learned inputs and outputs, as well as fresh inputs.
- **Learning to Learn:** The program develops its own inductive preference based on its prior experience with this type of problem.

6.2. SUPERVISED LEARNING APPROACH

In classification problems, supervised learning is quite popular because the purpose is to train a computer to study a classification system that has been built, which is quite common. A good example of this type of learning is digited recognition. Classification learning is appropriate for challenges in which the classification can be determined readily and in which presuming a classification is beneficial is assumed. Some situations where the mediator is able to resolve the classification problem on his or her own make it unnecessary to assign pre-determined classifications to each instance of a problem. In a classification context, this is an example of unsupervised learning, which may be seen here.

In supervised learning, there is a chance that the inputs will be left undefined on a number of occasions. If the necessary inputs are provided, this model is not required. In the absence of some input values, it is not possible to make any predictions regarding the outcomes of the simulation. Unsupervised learning is predicated on the assumption that all observations are begun by a latent variable, and it is predicted that the observations will be at the conclusion of the causal chain in this scenario. The examples of unsupervised and supervised learning are depicted in the diagram to the right (Figure 6.2).

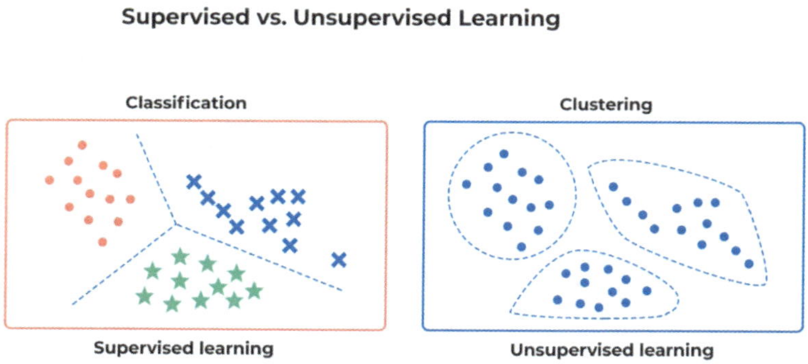

Figure 6.2. Illustration of supervised learning and unsupervised learning systems.

Source: https://www.intechopen.com/books/new-advances-in-machine-learning/types-of-machine-learning-algorithms.

The most widely used technique for training neural networks and decision trees is supervised learning, which is also the most effective. Both neural networks and decision trees rely heavily on the information provided by pre-determined classifications in order to function properly. As a result of the classification, errors in neural networks are determined. The classification then adjusts the network in order to decrease errors. In contrast, in decision trees, the qualities that provide the most information that aids in the solution of the classification enigma are decided by the classification process itself. Both of these strategies will be addressed in greater depth later on in this chapter. It is sufficient for knowledge to know that these examples thrive under "supervision" in the form of pre-determined categorization and that this is true (Figure 6.3).

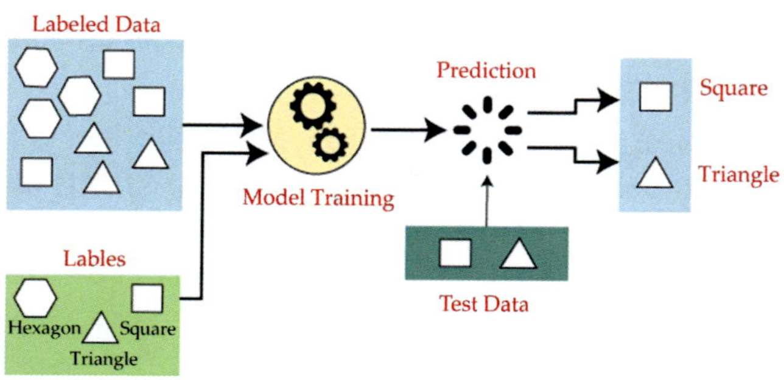

Figure 6.3. Supervised learning algorithm.

Source: https://geekycodes.in/what-is-supervised-learning/.

Inductive machine learning is the process of learning a set of rules from instances or, to put it another way, developing a classifier that aids in generalizing from fresh examples. The parts that follow describe how supervised machine learning is applied to a real-world situation. The initial phase entails gathering the data. If an essential specialist is available, she or he can advise on the most useful traits and features. If the expert is not available, a "brute-force algorithm" can be utilized, in which everything obtainable is measured in the hopes of making the appropriate features inaccessible. The dataset gathered using the "brute-force algorithm" is insufficient for induction. According to Zhang, it usually contains the needed information (2002). The data preparation and pre-processing are the next phase. Researchers have a variety of strategies to address missing data depending on the circumstances (Batista, 2003). Hodge (2004) presented a survey of current noise detection strategies. These researchers also shed insight on the benefits and drawbacks of these strategies. Instance selection is employed to deal with noise as well as the challenges of learning from huge datasets. In these datasets, the optimization challenge is instance selection, which aims to preserve mining quality while minimizing the sample size. It reduces the amount of data and uses a data mining method to work with enormous datasets effectively. For sampling the instances from the huge dataset, a variety of procedures are possible (Allix, 2000; Mooney, 2000).

In the field of feature subset selection, the method of identifying and deleting as many incorrect and discarded characteristics as possible is referred to as the selection of features (López, 2001; Lopez et al., 2002; Yu, 2004). As a result of this procedure, the dimensionality of data is reduced, allowing data mining algorithms to operate more quickly and efficiently than before. It is the fact that several features are reliant on one another that has a significant impact on the precision of supervised Machine language classification models, as previously stated. It is possible to find a solution to this challenge by developing creative features from a basic feature set of features. This technique is referred to as the feature construction technique. The recently developed features may prove useful in the development of classifiers that are more concise and exact. The finding of the relevant traits contributes to the enhanced clarity of the classifier that has been built, as well as the improved understanding of the notion. The implementation of Markov models and Bayesian networks in voice recognition is dependent on a few aspects of supervision in order to adjust the parameters in such a way that the mistakes may be minimized when given inputs are provided (Mostow, 1983; Fu and Lee, 2005; Ghahramani, 2008).

The most crucial thing to keep in mind is that the purpose of the learning process in a classification problem is to lower the error in accordance with the particular inputs being considered. These inputs, referred to as the "training set," are the specimens that aid the agent in its learning process. However, it is not always necessary to become familiar with these inputs. Suppose I want to demonstrate the exclusive-or and exhibit the combinations consisting of one true and the other false, for example (Getoor and Taskar, 2007; Rebentrost et al., 2014). The combination of both false and true is never presented; in this case, it is possible to learn the rule that states that the response is true at all times. Additionally, with machine learning algorithms, the problem of overfitting the data and basically remembering the training sets instead of adopting a broader categorization strategy is a problem that is commonly encountered (Rosenblatt, 1958; Durbin and Rumelhart, 1989). The inputs to each of the training sets are incorrectly categorized. This is a serious problem. This can result in issues if the implemented method is dominant enough to remember "special situations" that are not fit for the general principles of the algorithm in question. Overfitting can occur as a result of this. It is quite difficult to discover algorithms that are both powerful enough just to absorb complex functions and vigorous enough to yield outcomes that are generalizable over a wide range of situations (Figure 6.4) (Hopfield, 1982; Timothy, 1998).

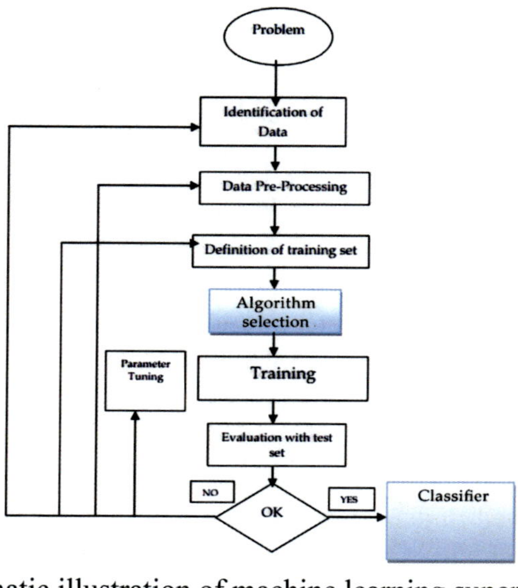

Figure 6.4. Schematic illustration of machine learning supervise procedure.

Source: https://www.intechopen.com/books/new-advances-in-machine-learning/types-of-machine-learning-algorithms.

6.3. UNSUPERVISED LEARNING

This strategy appears to be considerably more difficult: the purpose of this approach is to teach the computer how to do a task without our assistance. Unsupervised learning can be accomplished in two ways. The first way entails training the agent by implementing a reward system that shows achievement rather than by offering unambiguous categorizations. Because the purpose is to make decisions that will enhance the rewards rather than to produce a classification, this form of training is usually appropriate for the decision issue frame. This strategy is well-suited to the actual world, where agents may be rewarded for certain activities and chastised for others (LeCun et al., 1989; Bilmes, 1998; Alpaydm, 1999). Unsupervised learning is a type of supportive learning in which the agent's behaviors are based on previous punishments and rewards deprived of necessarily learning information about the direct ways in which they affect the real environment. This knowledge is useless in the sense that once the agent has been accustomed to the reward function, he or she knows exactly what action to take without having to think about it. The agent is well aware of the exact reward that must be obtained for each activity (Figure 6.5).

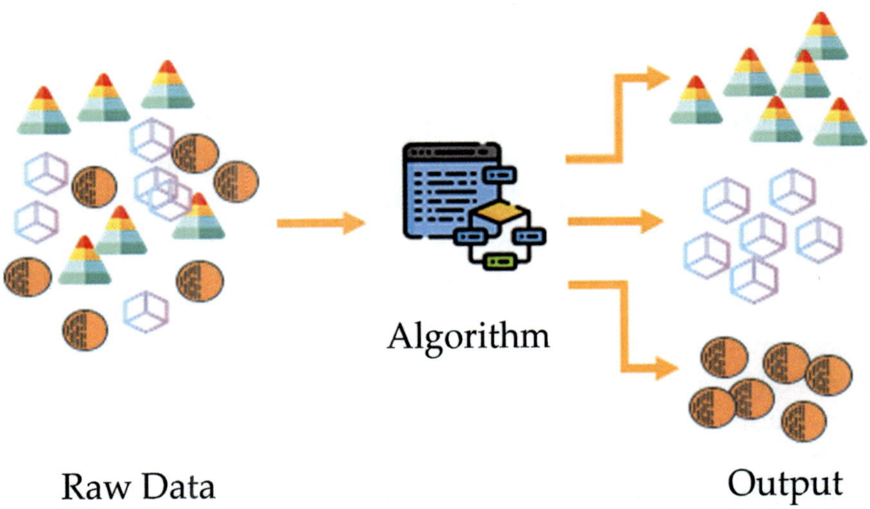

Figure 6.5. Illustration of unsupervised learning.

Source: https://medium.com/analytics-vidhya/beginners-guide-to-unsupervised-learning-76a575c4e942.

The use of this method can be particularly advantageous in situations where the calculation of each alternative takes an inordinate amount of time. However, learning via trial and error can be a very time-consuming process on the other hand. However, this sort of learning has the potential to be extremely effective because it does not rely on the classification of samples that have already been discovered. In some instances, our classifications are not the most accurate that could be achieved (Rumelhart et al., 1985, 1986; Schwenker and Trentin, 2014). Consider the case of backgammon, where the predictable knowledge in the game was turned on its head when computer programs that learned through unsupervised learning outperformed the best human chess players in terms of strength and endurance. There were specific ideas that these programs picked up that astounded the backgammon professionals, and they performed significantly better than the backgammon programs educated on the pre-classified samples. Clustering is another sort of unsupervised learning. The goal of this type of learning is to discover resemblances between training data and test data rather than to maximize the utility function (Xu and Jordan, 1996; Mitra et al., 2008).

Now, the assumption is that the clusters that have been discovered will correspond to the intuitive classification in a logical manner. For example, the clustering of persons based on demographics may result in the grouping of the wealthy in one set and the impoverished in the other set, depending on the circumstances. Despite the fact that the algorithm will not assign names to these clusters, it will be able to generate them and subsequently, by using these clusters, will be able to allocate fresh samples into one of the clusters (Herlihy, 1998; Gregan-Paxton, 2005). The data-driven technique described above is effective when a large amount of data is accessible; for example, social information filtering algorithms, which are employed by Amazon.com to propose books, are an example of such an approach (Stewart and Brown, 2004; Sakamoto et al., 2008). It is the concept of these algorithms that new customers are assigned to equivalent groups of people after they have been discovered and classified by them. It is enough for the algorithm to produce relevant results when social information filtering is used to gather information about other members of a cluster of people to be considered. In the remaining circumstances, clusters are simply a valuable tool for expert analysts to use in their analysis. Unfortunately, the unsupervised learning method is equally susceptible to the overfitting problem. There is no easy route to avoiding the problem of overfitting because the algorithm that can adapt to changes in its inputs must be powerful enough to overcome the problem (Nilsson, 1982; Vancouver, 1996; Sanchez, 1997).

Unsupervised learning methods are designed to remove structure from data samples without the assistance of a supervisor. Typically, the cost function is employed to determine the value of a structure, and it is minimized in order to derive the most favorable parameters that illustrate the hidden structure in data. The guarantee that the structures extracted are typical for the source is required for consistent and robust inference; for example, the structures extracted from the second model set of a comparable data source must be the same as those retrieved from the first model set of a similar data source (Bateson, 1960; Campbell, 1976; Kandjani et al., 2013). Overfitting is a term used in the statistical and machine learning literature to describe a lack of robustness in a model. Currently, the overfitting phenomena are described by a group of histogram clustering dummies, which play a key role in information recovery, linguistics, and computer vision applications, among other things. Because of the enormous deviations in the outcomes, learning algorithms with the potential to simulate fluctuations have emerged as the ultimate entropy principle for the process of learning (Pollack, 1989; Pickering, 2002; Turnbull, 2002).

Many successes have been produced by unsupervised learning, such as:

- World champion-caliber backgammon program;
- Machines skilled in driving cars.

When there is a simple way to assign values to activities, unsupervised learning can be a powerful tool. Clustering is helpful when there is enough data to construct clusters, and especially when supplementary data about a cluster's adherents are used to generate additional outcomes due to data dependencies (Acuna and Rodriguez, 2004; Farhangfar et al., 2008). When it is known whether the classifications are true or are just random objects that the computer can recognize, classification learning is useful. Classification learning is typically required in scenarios when the algorithm's decision is required as input in another context. Clustering and classification learning are both beneficial, and the best strategy to use depends on the following factors (Dutton and Starbuck, 1971; Lakshminarayan et al., 1999):

- Kind of a problem being solved;
- Time allotted for solving it;
- Either supervised learning is possible.

6.4. ALGORITHM TYPES

In the extent of supervised learning where which mostly classification is dealt with, the following are the types of algorithms:

- Linear classifiers;
- Naïve bayes classifier;
- Logical regression;
- Support vector machine;
- Quadratic classifiers;
- Perceptron;
- Boosting;
- Decision tree;
- K-means clustering;
- Neural networks;
- Random forest;
- Bayesian networks.

6.4.1. Linear Classifiers

Classification is used in the machine learning process to arrange objects with similar feature values into groups. As per Timothy et al. (1998), linear classifiers achieve this with the help of better decisions (Grzymala-Busse and Hu, 2000; Grzymala-Busse et al., 2005). The categorization decision is influenced by the linear combination input. The output is supplied as if the real vector $\vec{x}$ were the input to the classifier:

$$y = f(\vec{w} \cdot \vec{x}) = f\left(\sum_j w_j x_j\right),$$

where; $\vec{w}$ → real weights vector, f → function in which the dot product of two vectors is translated into the preferred output. A set of marked training samples helps to deduce the vector $\vec{w}$. The function f is usually used to translate values over a certain threshold to first class and the remainder of the data to second class. The complex function f determines the likelihood that an object belongs to a specific category (Figure 6.6) (Li et al., 2004; Honghai et al., 2005; Luengo et al., 2012).

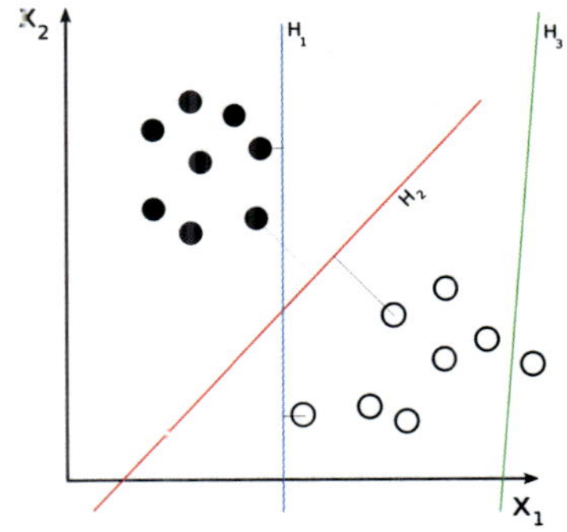

Figure 6.6. Graphical representation of linear classifiers.

Source: https://en.wikipedia.org/wiki/Linear_classifier.

Note: Any number of linear classifiers can correctly classify the solid and empty dots in this example. H_1 (blue) and H_2 both appropriately classify them (red). H_2 is "better" in the sense that it is the furthest away from both groupings. The dots are incorrectly classified by H_3 (green).

The linear classifier's functionality in two-class classification can be viewed as dividing the high-dimensional input vector with a hyperplane. The points on one side of the hyperplane are labeled "yes," while those on the other are labeled "no." Because linear classifiers are the fastest classifiers, especially when the real vector is sparse, they are frequently utilized in situations where classification speed is a concern. Decision trees can also be made to run quicker (Hornik et al., 1989; Schaffalitzky and Zisserman, 2004). When the number of dimensions in the real vector is big, linear classifiers typically perform well. Every element in the real vector in document categorization is usually the count of a specific word in the document. In these situations, the classifier must be well-regularized (Dempster et al., 1997; Haykin and Network, 2004).

According to (Luis et al.), a support vector machine (SVM) performs classification by building an N-dimensional hyperplane that divides data into categories as efficiently as possible. SVM models and neural networks are inextricably linked. The sigmoid kernel function is utilized in the SVM model, which is similar to the two-layer (PNN) perceptron neural network (Rosenblatt, 1961; Dutra da Silva et al., 2011).

Traditional multilayer (PNN) perceptron neural networks are closely connected to SVM models. SVM models, which use a kernel function as a training scheme, are a good alternative to polynomials, radial basis functions, and multilayer perceptron neural networks classifiers. Instead of tackling a non-convex and unconstrained minimization issue like in normal neural network instruction, the weight of the network is determined by solving a quadratic programming problem with linear constraints in SVM (Gross et al., 1969; Zoltan-Csaba et al., 2011; Olga et al., 2015). The attribute is a predictor variable in the SVM literature, and a feature is a changed attribute that specifies the (HP) hyperplane. Feature selection is the process of selecting the best appropriate representation. A vector is a collection of features that describe one of the situations. The purpose of SVM modeling is to find the best hyperplane for splitting vector groups so that cases belonging to one category of the target variable are on one side of the plane and cases belonging to the other category are on the other (Block et al., 1962; Novikoff, 1963; Freund and Schapire, 1999). The support vectors are the vectors that are nearest to the hyperplane. The next section gives an overview of the SVM procedure.

6.4.2. A Two-Dimensional Case

Before we look at N-dimensional hyperplanes, let's have a look at a simple 2-dimensional illustration. Consider the following scenario: we want to perform classification, and the data we have available has a specified target variable with two categories. Also, consider the possibility that two predictor variables with continuous values are available as well. If the data points are plotted on the X and Y axes with the values of the predictor on the X and Y axes, the image displayed below may be the result. The rectangles indicate one category, and the ovals represent the other category, as shown in Figure 6.7 (Lula, 2000; Otair and Salameh, 2004; Minsky and Papert, 2017).

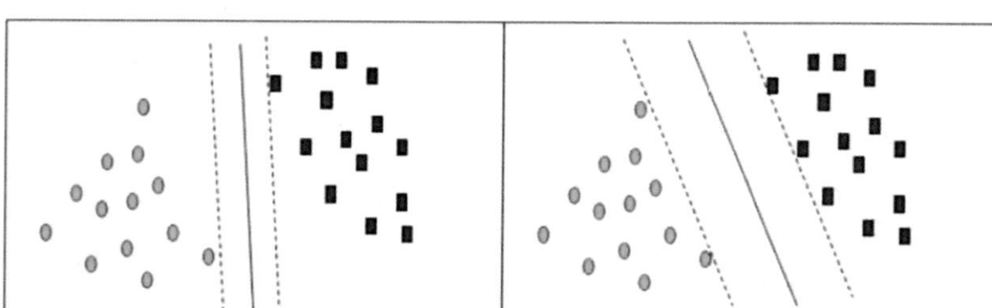

Figure 6.7. The demonstration of SVM analysis for determining 1D hyperplane (i.e., line) which differentiates the cases because of their target categories.

Source: https://www.intechopen.com/books/new-advances-in-machine-learning/types-of-machine-learning-algorithms.

In this instance, the cases of one category are in the lower-left corner, while the examples of another category are in the upper right corner. Both scenarios are completely distinct from one another. The one-dimensional hyperplane that separates both cases on the basis of their target categories is sought by the SVM analysis. There are an infinite number of alternative lines; the two candidate lines are depicted above. Now the question is, which line is far superior, and how is the optimum line defined? (Caudill and Butler, 1993; Hastie et al., 2009; Noguchi and Nagasawa, 2014).

The dotted lines parallel to the dividing line indicate the distance between the dividing line and the nearest vectors to the line. The space between dotted lines is the margin. The points that define the margin's size are known as support vectors. It is depicted in the diagram below.

The line, or, in particular, the hyperplane that is angled to maximize the margin among the support vectors, is discovered using the SVM analysis (Figure 6.8) (Luis, 2005; Hall et al., 2009).

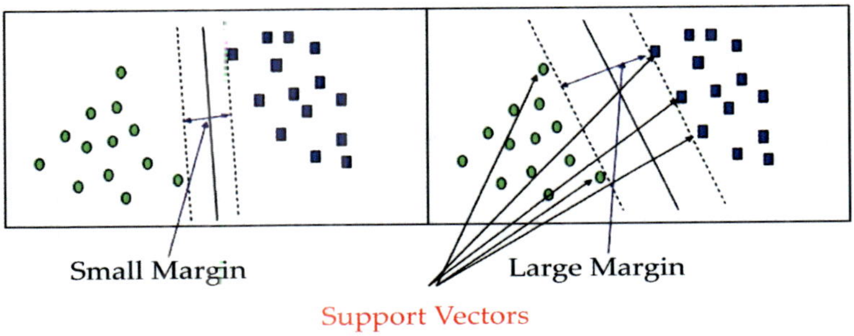

Figure 6.8. Illustration of an SVM analysis containing dual-category target variables possessing two predictor variables having the likelihood of the division for point clusters.

Source: https://www.intechopen.com/books/new-advances-in-machine-learning/types-of-machine-learning-algorithms.

The line in the right section is superior to the line in the other section, as shown in the diagram above. Life would be stress-free if all analyzes included two category goal variables, two predictor variables, and a straight line that could separate the group of points (Neymark et al., 1970; Hammes and Wieland, 2012). Unfortunately, this is not usually the case; therefore, the SVM must deal with:

- Handling of the cases where more than the two predictor variables distinct the points with the non-linear curves.
- Handling of the cases where groups cannot be separated completely.
- Handling of the classifications having more than the two categories.

Three major machine learning techniques are explained in this chapter with examples and how these techniques perform in reality. These are:

- K-means clustering;
- Neural network;
- Self-organized map.

6.4.3. K-Means Clustering

This technique consists of a few simple steps. In the beginning, K (the cluster number) is established, and the cluster center is assumed. The primary center might be any random item, or the first K entities in the structure can alternatively serve as the primary center (Teather, 2006; Yusupov, 2007). The K means algorithm then performs the three steps listed below till convergence. Iterate until the result is stable.

- Determining the center coordinate;
- Determining the distance of every object from the center;
- Grouping the objects based on the minimum distance.

K-Means flowchart is shown in Figure 6.9. K-means clustering is the simplest unsupervised learning technique that may be used to solve the well-known segmentation problem. It is also the most widely used algorithm. The procedure uses a straightforward approach to categorize the given data set into one of a pre-determined number of clusters. It is necessary to define K centroid values for each cluster. This is important since different results can be achieved from different locations (Franc and Hlavá, 2005; González et al., 2006). These centroids must be placed in a strategic manner. The more appropriate option is to place the centroids as far apart as possible from one another. It is next necessary to take every point from the provided data and associate it with the centroid nearest to that point in the data set. When there are no more points, it signifies that the first stage has been completed, and an initial grouping is carried out (Oltean and Dumitrescu, 2004; Fukunaga, 2008; Chali, 2009).

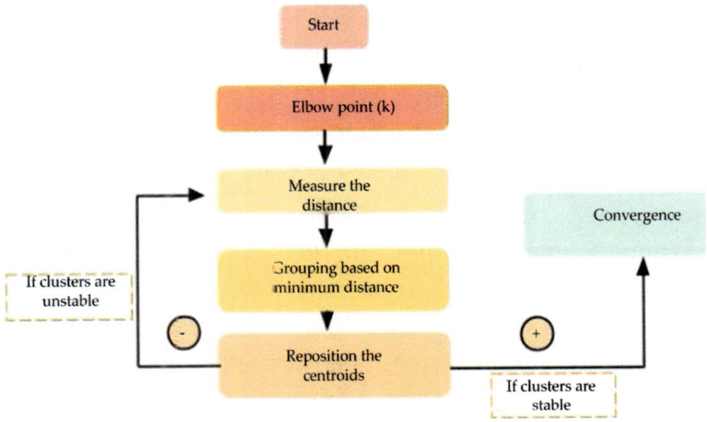

Figure 6.9. Schematic illustration of K-means iteration.

Source: https://www.intechopen.com/books/new-advances-in-machine-learning/types-of-machine-learning-algorithms.

It is now necessary to recalculate 'knew centroids. After acquiring the new centroids, a new binding must be performed between the same points and the nearby new centroid. The result is a loop. The k centroids gradually shift their position because of this loop because there are no more changes. As a result, the centroids stop moving (Oltean and Groşan, 2003; Oltean, 2007). This algorithm aims to reduce an objective function, in this case, a squared error function. The role of the objective:

$$J = \sum_{j-1}^{k} \sum_{i-1}^{x} \left\| x_i^{(j)} - c_j \right\|^2$$

$\left\| x_i^{(j)} - c_j \right\|^2$ It is a selected distance measure between $x_i^{(j)}$ (data point) and C_j (cluster center). It is a sign of distance of *n* data points with respect to their particular cluster centers.

The algorithm for this technique consists of the steps defined below:

- K points are placed into the space denoted by objects which are being clustered. These points denote primary group centroids;
- Each object is assigned to the group which has the nearest centroid;

- The positions of K centroids are recalculated after all of the objects have been allocated;
- Steps ii and iii are repeated as long as the centroids are moving. This creates a partition of objects into the groups from where the minimized metric can be calculated.

Despite the fact that it can be determined that the process will always finish, the k-means clustering algorithm does not fundamentally identify the most optimal configuration that is identical to the minimum global objective function in most cases (Burke et al., 2006; Bader-El-Den and Poli, 2007). It is also important to note that the K-means method is extremely sensitive to the major cluster centers that are arbitrarily selected. This procedure is repeated multiple times in order to reduce the impact of this effect. The K-means method has been adapted to a wide range of problem domains. When it comes to working with fuzzy feature vectors, this approach is a strong contender for modification (Tavares et al., 2004; Keller and Poli, 2007).

Suppose that n-sample feature vectors (x_1, x_2, x_3, …, x_n) are available, all of which belong to the same class, and that they all fall into one of the k dense clusters where k n is the number of samples available. Assume that mi represents the mean of the cluster i. Because they are well separated, we can use the minimal distance classifier to divide the clusters and use it to split the data. It is possible to say that the x is included in cluster I only if || x – mi || represents the shortest of all k distances between the two points (Figure 6.10). This recommends the following process in order to find the k means:

- Supposing the values for means (m_1, m_2, m_3, …, M_K) unless no modifications in any of the mean occur;
- Classifying the samples into the clusters using the estimated means;
- For the cluster **i** from the range 1 to k;
- Swap **mi** by mean of all the samples for the cluster **i**;
- end_for;
- end_until.

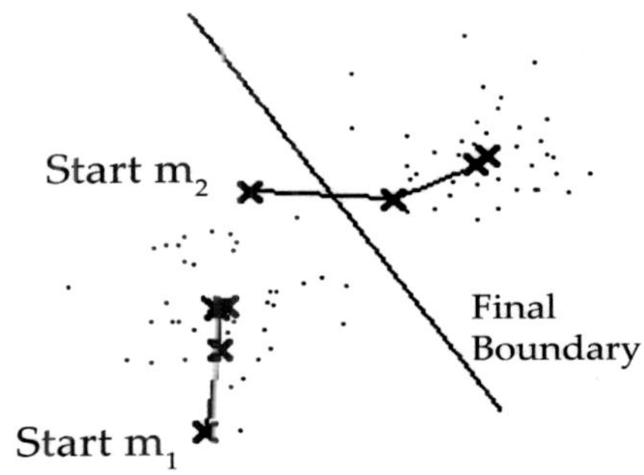

Figure 6.10. Demonstration of the motion for m_1 and m_2 means at the midpoint of two clusters.

Source: https://www.intechopen.com/books/new-advances-in-machine-learning/types-of-machine-learning-algorithms.

This is a simplified version of the k-means algorithm. It can be thought of as a greedy technique for grouping n models into k clusters in order to minimize the total number of squared distances between cluster centers. There are certain flaws in this algorithm:

- The procedure for determining the means was not specified. One well-known method for starting the mean is to select k of the models at random.
- The results are determined by the fundamental mean values, and it is usual for suboptimal partitions to emerge. The conventional technique is to choose a variety of different beginning positions.
- It's conceivable that a set of models close to mi is empty, preventing mi from being updated. This is an annoyance that must be addressed during implementation.
- The results are influenced by the measure used to quantify || x – mi ||. The traditional solution is to normalize each variable by its standard deviation, albeit this is not always desired.

6.4.4. Neural Network

In fact, neural networks may perform multiple reversal and classification tasks at the same time, albeit they usually only do one at a time (Forsyth, 1990; Bishop, 1995; Poli et al., 2007). In most instances, the network's output variables are limited to just one. This may be related to the number of outcome units in several state classification issues. If you draw a network with a lot of output variables, crosstalk is a possibility. Hidden neurons have a hard time learning since they are trying to simulate at least two functions. The optimum way is to train a separate network for each output and then integrate them into one so that they can operate as a unit. Further subsections explain how to use neural networks.

6.4.4.1. Multilayer Perceptron

In today's world, this is the most widely acknowledged network architecture, having been discovered by Rumelhart and McClelland (1986) and thoroughly detailed in most neural network textbooks. It is briefly mentioned in the preceding sections that this type of network has the following characteristics: the units, in turn, implement the weighted sum of the inputs and, with the help of a transfer function, pass through this activation level in order to generate the output. Each of the units is configured in a covered feedback control topology. The form of the input-output model, as well as the weights of the multilayer perceptron, can be explained in a straightforward manner. As a result, the free variables in the model are biased by this network. By varying the number of layers and the number of units in each layer, such networks may be used to construct functions of random complexity, with the complexity of the function determined by the number of levels and the number of units in each layer. When designing MLPs (multilayer perceptrons), one of the essential problems to consider is the specification of hidden nodes as well as the number of units included in these hidden levels (Figure 6.11) (Bishop, 1995; Michie, 1994; Anevski et al., 2013).

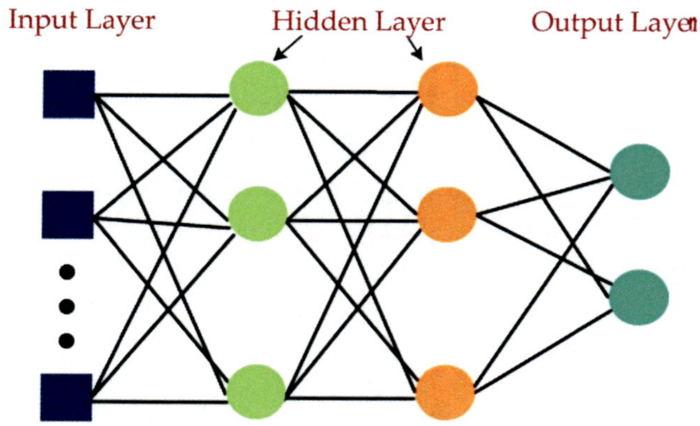

Figure 6.11. Example of multi-layer perceptron in TensorFlow.

Source: https://www.javatpoint.com/multi-layer-perceptron-in-tensorflow.

The problem interprets the total number of input-output units. There may be some ambiguity about which inputs to use, which will be explained later. Let's pretend for the time being that the inputs are chosen intuitively and are significant. The number of hidden units to be used is obvious. Using a single hidden layer with several units equal to half of the total number of input-output units can be a suitable place to start. Later, we'll go through how to pick an acceptable amount (Nilsson, 1982; Orlitsky et al., 2006).

6.4.4.2. Training Multilayer Perceptron

The weights and criteria of the network must be established after the number of layers and units inside each layer have been determined to reduce the network's prediction error. This is done through the training algorithms. The historical cases are used to automatically control the weights and thresholds to reduce the mistake (Mohammad and Mahmoud, 2014). This approach is identical to fitting the network-defined model to the provided training data. Running all the training instances throughout the network and comparing the real output with the desired outputs can reveal the individual configuration's fault (Ayodele, 2010; Singh and Lal, 2013). With the use of an error function, the differences are pooled together to form the network error. The most well-known error functions are the following:

- For regression situations, the sum squared error is employed. The discrete errors of the output nodes are squared and then added together as a total squared error;
- Cross-entropy functions are used for the maximum likelihood classification.

In traditional modeling approaches, such as linear modeling, it is likely that the design of the model will be determined algorithmically to decrease this mistake to an absolute minimum. The trade-off made in exchange for the non-linear modeling power of neural networks is that we can never be guaranteed that the error cannot be minimized any longer under any conditions (Jayaraman et al., 2010; Punitha et al., 2014).

The concept of error surface is one that is interesting to explore in this context. It is necessary to engage each one of the N weights and network starting points in order to be the dimension in space. The network error is often represented as the N + 1th dimension. Any possible configuration of the weights can result in the error being intrigued in the N + 1th dimension, resulting in the formation of an error surface. Network training is primarily concerned with finding the lowest point on a multidimensional surface, which is the primary goal. The error surface of a linear model with a sum squared error function is a parabola in the case of the sum squared error function. In other words, the error surface is shaped like a bowl with a single minimum value. As a result, determining the bare minimum is straightforward (Kim and Park, 2009; Chen et al., 2011; Anuradha and Velmurugan, 2014).

In the case of neural networks, the error surfaces are more complicated and are frequently characterized by a large number of non-supportive properties such as:

- Local minima;
- Flat-spot and plateau;
- Saddle-points;
- Long narrow ravine.

As a result, it is unlikely that an analytical determination can be made as to where the general minimum of error surface is located; hence, the training of neural networks involves an exploration of the error surface. The training algorithms attempt to find the global minimum step by step, starting with a basically arbitrary setup of the weights and thresholds and progressing from there. Typically, the gradient of the error surface is considered at the most recent position and is then employed to make the downward maneuver. A final stop is reached by the algorithm at the lowest position, which may be either the local minimum or the global minimum (Fisher, 1987; Lebowitz, 1987; Gennari et al., 1988).

6.4.4.3. Back Propagation Algorithm

The backpropagation technique is the best example of neural network training (Haykin, 1994; Patterson, 1996; Fausett, 1994). Conjugate gradient descent and the Levenberg-Marquardt (Bishop, 1995; Shepherd, 1997) are two contemporary second-order algorithms that are significantly faster to use in many tasks. The backpropagation method still offers numerous advantages in various scenarios and can be the easiest algorithm to understand. This algorithm will only be introduced; more complex algorithms will be explored later. The slope vector is calculated in the backpropagation technique. The slope vector from the present position points to the steepest descent line. The mistake will be reduced if we move a modest distance along the slope vector. A series of such maneuvers will eventually yield a minimum of some sort. The difficult part is deciding how big the step should be (Figure 6.12) (Michalski and Stepp, 1982; Stepp and Michalski, 1986).

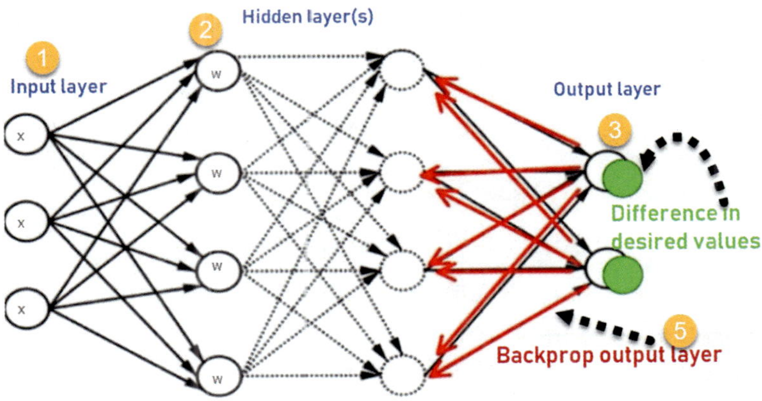

Figure 6.12. Backpropagation algorithm working flowchart.

Source: https://www.guru99.com/backpropogation-neural-network.html.

Large steps may be taken fast, yet these steps may be in excess of the solution or on the incorrect route altogether. One illustration of this is the circumstance where the algorithm goes extremely slowly over a steep and narrow valley, reflecting from one side to the other. Small steps, as compared to large leaps, may move the needle in the right direction, but these steps will require a number of rounds. The step size is proportional to the slope and the rate of acquisition. The specific setting for the particular constant learning rate is always dependent on the application and is typically determined by the experimenter during the experiment. It is also possible that the learning rate is time-varying and that it decreases with the expansion of the algorithm (Knill and Richards, 1996; Mamassian et al., 2002).

The addition of a momentum component to the algorithm is widely used to increase its performance. This accelerates the travel in a specific direction, and if a sufficient number of steps are taken in that direction, the algorithm becomes more efficient. It is because of this speed that the algorithm is capable of escaping the local minimum on occasion, as well as moving fast over the flat patch and plateau (Weiss and Fleet, 2002; Purves and Lotto, 2003).

As a result, the algorithm iteratively repeats the process a number of times. Every single epoch, the cases yield to the network and the goal, creating a chance for them both. After that, the actual outputs are compared, and the error is computed. When combined with error surface gradient, this error is used to adjust the weights, after which the procedure is repeated.

The primary network configuration is chosen at random, and the training is terminated once a pre-determined number of epochs have passed or after the error has reached a tolerable threshold. Whenever an error is halted in order to improve, the training is also stopped (Yang and Purves, 2004; Rao, 2005; Doya, 2007).

6.4.4.4. Over-Learning and Generalization

One of the major problems with the method described above is that the backpropagation algorithm does not actually reduce the error in which we are truly interested, which is the consistent error that the network will generate when the new cases succumb to it. This is one of the most serious shortcomings of the method described above. The ability of the network to simplify new cases is the most important quality it possesses. While the network is capable of reducing error, it is lacking in the area of a perfect and extremely big training set. This is not the same as the diminishing of an error on the true error surface, which is the error of the basic and unfamiliar model; instead, it is the opposite (Figure 6.13) (Dockens, 1979; Bishop, 1995; Abraham, 2004).

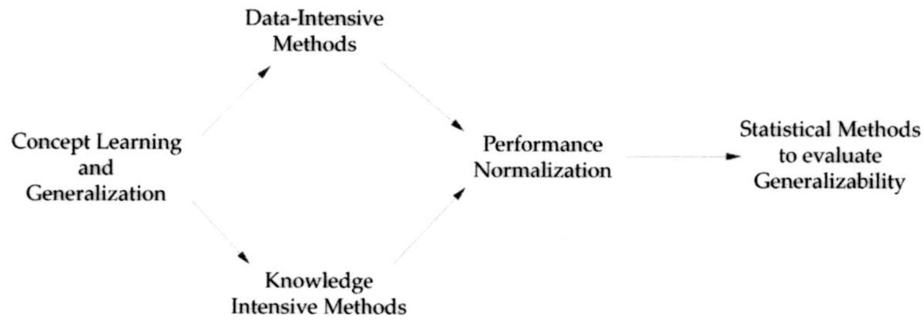

Figure 6.13. The relationship between concept learning, generalization, and generalizability.

Source: https://www.researchgate.net/figure/The-relationship-between-concept-learning-generalization-and-generalizability_fig1_3296799.

In this regard, the over-learning problem serves as the most striking illustration of the distinction. It is relatively simple to validate this concept using polynomial curve fitting rather than neural networks, as opposed to neural networks. A polynomial is an equation that consists solely of constants and the power of the variables in the equation.

For example:

$y = 3x_2 + 4x + 1$

Different polynomials will always have a variety of shapes to them. The shapes of the polynomials with larger powers become increasingly bizarre as the number of powers increases. It is possible that we will need to fit the polynomial curve to a specific collection of data to characterize the data. Because the data is likely to be noisy, we should not expect the best possible model to pass through all of the points in an exact manner (Long, 1980; Konopka, 2006; Shmailov, 2016). It is possible that the low-order polynomial is not sufficiently flexible to suit the points in close proximity. High-order polynomials are essentially overly supple in comparison to lower-order polynomials, fitting the data perfectly by implementing a very eccentric shape that is not linked to the core function (Figure 6.14) (Seising and Tabacchi, 2013; Shmailov, 2016).

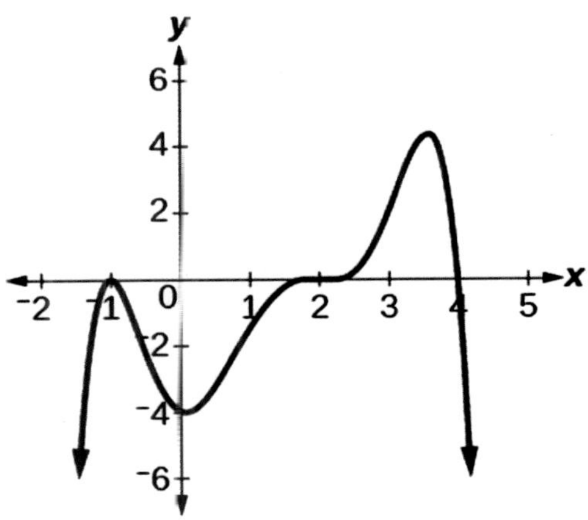

Figure 6.14. Graph showing a typical polynomial function.

Source: https://courses.lumenlearning.com/wmopen-collegealgebra/chapter/graphs-of-polynomial-functions/.

The neural networks, in particular, face the same issue. The one with higher weights represents a more complex function and therefore is more prone to over-fitting. The one with less weight, on the other hand, may not be efficient enough to mimic the core function. The network with no hidden layers, for example, essentially simulates a more straightforward linear function. So, how can we choose the network's precise complexity? The larger network will almost always achieve the lower error in the end, although this could indicate over-fitting rather than effective modeling (Somorjai et al., 2003; Zanibbi and Blostein, 2012).

The answer to the above question is to compare the development to an independent data collection. In the backpropagation process, certain examples are kept and not really used for training. Instead, they're used to keep an independent eye on the algorithm's progress (Lukasiak et al., 2007; Puterman, 2014). It is always the case that the network's principal execution on the training and independent selection sets is the same. As the training progresses, the error decreases naturally, and as the error function decreases, the selection error decreases as well (Ahangi et al., 2013; Aldape-Pérez et al., 2015).

If, on the other hand, the selection error begins to increase or stops lowering, this indicates that the network is beginning to overfit the available data, and training should be stopped. Over-learning is what happens when overfitting occurs throughout the training process, as described above. Because the network is powerful enough to handle the problem, it is usually desirable to lower the number of hidden layers in this circumstance. Instead, if the network isn't powerful enough to model the fundamental function, there isn't likely to be any over-learning, and neither the training nor the selection mistakes will be acceptable (Goodacre et al., 2007; Brereton, 2015).

Because of the issues with local minima and the judgments about the size of the network to use, using a neural network usually entails experimenting with a large number of different networks. Each of the networks was trained many times, and the performance of each network was monitored separately (Sutton, 1988; Sutton and Barto, 1998). The major guidance for better performance here is selection error. However, in accordance with the standard scientific premise that all other things are equal, a straightforward and simple model is always chosen over a complex model. With a negligible improvement in the selection error, a smaller network can be chosen over a bigger network (Ambikairajah et al., 1993; Russell and Norvig, 2016).

The trouble with this approach of recurrent study is that the selection set plays the most important role in determining the model, implying that it is essentially a stage in the training phase. As a result, its dependability as a self-contained performance guide for the model is jeopardized. With the help of suitable experiments, a network that functions properly could be discovered. Retaining the third set of instances, referred to as the test set, is a common practice to ensure the final model's performance. The final model is next validated using test set data to ensure that the training and selection set outcomes are genuine and not training procedure elements. The test set must be utilized once to properly justify its role; if it is used to alter and recapitulate the training phase, it then becomes the selection data (Hattori, 1992; Forseth et al., 1995).

Even for a single subset, this partitioning into numerous subsets is unlucky since we frequently have less data than we would have preferred ideally. By resampling, we can find a solution to this difficulty. Selection and test set experiments can be arranged using multiple divisions of the data provided in the training set (Gong and Haton, 1992; Wagh, 1994). The numbers of methods to this subset are:

- Random resampling;
- Cross-validation;
- Bootstrap.

If we base design decisions on tests with diverse subsets of examples, such as using the highest suitable configuration of the neural network, the results will be more reliable. We can use those experiments exclusively to:

- Lead the decision as to use which of the network types and teach such networks from very scratch with the help of new samples;
- Preserve the best possible networks originated during the process of sampling.

To summarize, a network design obeys the following steps:

- Selection of an initial configuration;
- Conducting a series of tests with each configuration iteratively and then maintaining the best network discovered. If the training identifies a local minimum, a series of tests for each configuration is required to avoid being duped. It's beneficial to resample;
- On each test, if under-learning befalls, then try adding extra neurons to the hidden layer. If this isn't useful, try adding one more hidden layer;
- If over-learning befalls, then try removing the hidden units;
- Once an efficient configuration for the networks is determined, resample, and produce new networks with the help of that particular configuration.

6.4.4.5. Selection of Data

All of the processes listed above are predicated on a single premise. The training and test data should both be representative of the core model, to put it another way. One of the most famous computer science proverbs, "garbage in, garbage out," couldn't be more applicable than in the case of neural modeling. If the training data is not representative of the real world, the value of the model is diminished. The worst-case scenario is that it is completely ineffective. It is extremely helpful in showing the types of difficulties that can cause the training set to be altered.

The training data is almost often historical in nature. If the circumstances have changed, it is possible that relationships that existed in the past will no longer exist. It is necessary to consider all of the possibilities. The scenarios that are presented can only aid in the learning process of the neural network. If people who earn more than $80,000 per year are at risk of having bad credit, and the training data includes no one who earns more than $40,000 per year, it is unlikely that the neural network will make the correct decision if it comes across one of the previously undetected cases later in the training process. Extrapolation is risky with any sort of network, but certain categories of neural networks may perform particularly poorly in such circumstances due to their inherent complexity.

A network is efficient at absorbing properties that are easy to learn. A typical instance of this principle is a project that was developed to recognize tanks automatically. Hundreds of photographs with tanks on them, as well as hundreds of pictures without tanks, are used to train a network. It achieves a perfect 100% score in the final exam. When the results are compared to the new data, they are completely inconclusive. In this case, it is because the photographs with tanks on them were shot on gloomy and rainy days, whereas the network absorbs enough light to distinguish the differences between the pictures taken in bright sunlight. In order for this network to function properly, training scenarios would need to be developed that encompasses all of the weather and illumination conditions under which the network is expected to operate.

Because a network minimizes the chance of mistakes, the proportion of different types of data in the collection is crucial. If a network is trained on a collection of data that contains 900 good cases and 100 bad cases, it will bias its results in favor of the good cases because doing so will allow the algorithm to minimize the overall error. If the depiction of the scenarios in the real population differs from the depiction in the network, the conclusion made by the network may be incorrect. An excellent illustration of this is the process of diagnosing a disease. Consider the possibility that 90% of the people who are typically screened are devoid of any ailment. A network is skilled with a spilled of 90/10 based on the data set that is accessible. This network is then used on individuals who are complaining of specific difficulties and where the likelihood of the existence of any disease is 50/50, such as in the case of diabetes. In other circumstances, the network will behave cautiously and will be unable to diagnose the disease because of this. In contrast, if the network is trained on "complainants" data and then explored on "regular" data, it is possible that the network will generate a substantial number of false positives. In such circumstances, a data set may be produced to account for the circulation of data, or the decisions made by the network may be adjusted because of the inclusion of a loss matrix into the network (Bishop, 1995). Frequently, the most effective strategy is to ensure that all instances are presented in an equal manner and then to explain the choices in the appropriate light.

6.4.5. Self-Organized Map

When compared to other networks, SOFM (self-organizing feature map) networks are used in a different manner. Other networks are being planned for supervised learning tasks, but the SOFM networks are being planned primarily for unsupervised learning tasks, as previously stated. Haykin (1994), Patterson (1995), and others (Fausett, 1994). In supervised learning, the training data set consists of examples having input variables that are organized with the corresponding outputs, whereas in unsupervised learning, the training data set consists solely of input variables and no outcomes. At first glance, this may appear to be a strange arrangement. What can a network learn if its outputs aren't there? According to the answer to this question, the SOFM network makes an attempt to absorb data configurations by itself (Figure 6.15).

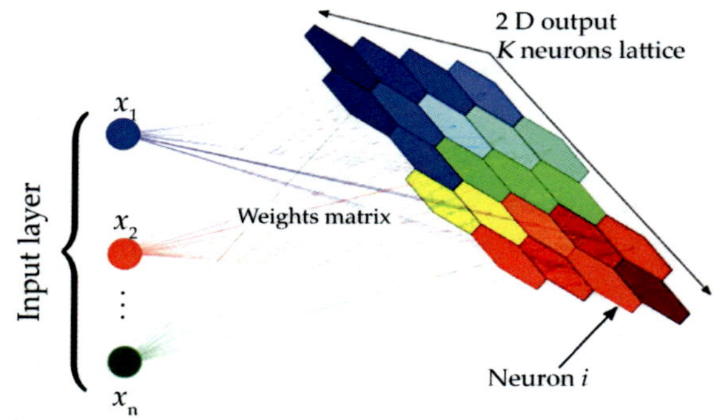

Figure 6.15. Illustration of self-organized maps.

Source: https://www.superdatascience.com/blogs/self-organizing-maps-soms-how-do-self-organizing-maps-work.

According to Kohonen (1997), one conceivable application of SOFM is in the exploratory data evaluation process. The SOFM network can be used to locate data clusters and to link classes that are similar to one another. Users of the network can determine the insight into the data, which is subsequently used to improve the network as a result of the consumer's efforts. As soon as the classes of data are identified, they can be labeled in order to enable the network to perform the classification tasks that have been assigned to it. It is possible to employ SOFM networks for classification when the outcome classes are available immediately after the network is created. The advantage they have in this case is their ability to draw attention to the similarities that exist across the classes.

A second potential application for this network is in the detection of novel objects. It is possible for the SOFM networks to learn to recognize clusters in training data and then respond to them. If new data is discovered and the network is unable to identify it, this indicates that the data is novel.

The SOFM network consists of the following two layers:

- The input layer; and
- An output layer with radial units (topological map layer). In the topological map layer, the units are positioned in space, i.e., in the two dimensions.

SOFM networks are trained using an iterative approach (Patterson, 1996). The technique starts with a random set of radial centers and gradually modifies them to expose the data clustering. At one point, this technique is linked to subsampling and the K-Means algorithms, both of which are used to distribute centers in the SOM (Self-Organize Map) network. To allocate centers for these types of networks, the SOFM algorithm might be used.

The iterative training technique additionally organizes the network so that the units displaying centers near one other in the input layer are also near each other on the topological layer. The network's topological layer is a rudimentary 2-dimensional lattice that must be crumpled and slanted into an N-dimensional input layer in order to maintain the original structure as much as feasible. Any attempt to reduce an N-dimensional space to two dimensions will result in a loss of detail. This strategy can be beneficial in allowing the user to think about data that would otherwise be difficult to comprehend.

The fundamental iterative SOFM method cycles over a number of periods, with each epoch applying the training example and utilizing the algorithm below:

- Selection of the winning neuron. The neuron whose center is closest to the input situation is known as the winning neuron;
- We are adjusting the winning neuron so that it can resemble the input case.

The time decaying learning rate is used in the iterative technique to achieve the weighted sum. It also shows that as time passes, the changes become more sensitive. This ensures that the centers unite in a cooperative display of the cases that aid the specific neuron's victory. The topological assembly property is achieved by incorporating the concept of the neighborhood into the iterative method. A neighborhood is a group of neurons that surround the winning neuron. Similar to the learning rate, the neighborhood deteriorates over time. Initially, the neighborhood has a large number of neurons, but as time passes, the area will become vacant. In the Kohonen or SOFM method, all of the neighbors in the current neighborhood are subjected to neuron change.

As a result of these neighborhood apprises, large regions of a network are largely drawn towards the training examples and dragged significantly. Similar cases stimulate clusters of neurons within the topological map, and the network cultivates a rudimentary topological organization. With the passage of time, both the learning rate and the neighborhood shrink, allowing for better distinctions within the map's areas, eventually leading to the fine-tuning of a single neuron. Frequently, the instruction is split into two sections:

- A comparatively short phase with usually high learning rates and neighborhood; and
- A long phase has low learning rates and zero neighborhoods.

Once the network has been successful in identifying configuration in the data, it may be used as a visualization device to examine the data. With the assistance of this network Win, Frequencies Datasheet may be studied in order to determine whether or not separate clusters have been formed on the map of the world. It is necessary to implement individual scenarios and evaluate the topological map in order to determine whether or not any importance can be assigned to clusters. Once the clusters have been identified, the neurons included in the topological chart are labeled in order to describe the significance of the neurons in the cluster. When the topological chart is displayed in this manner, it becomes possible to introduce new situations into the network. The network is capable of executing an arrangement under the condition that the winning neuron is tagged with the appropriate class name. If this is not the case, the network is classified as undecided.

When doing classification, SOFM networks make use of the accept threshold as well as they reject threshold. The distance between the input and the triggering extent of a neuron in a SOFM network determines its triggering extent, and the accept threshold determines the maximum documented distance. If the initiation of the winning neuron is much more than the distance between the two neurons, the SOFM network is regarded to be uncertain. In order to operate as the novelty detector, a SOFM network must be properly labeled with tags and with the accept threshold appropriately assigned.

Several well-known properties of the brain are used to drive Kohonen's SOFM network (Kohonen, 1997). When you think about it, the cerebral cortex is basically just a big smooth sheet with certain distinct topological qualities.

6.4.5.1. Grouping Data via Self-Organized Map

The data represents the first section of the Self-Organize Map. Below are a few samples of three-dimensional data commonly used in Self-Organize Maps testing. The colors are now shown in three dimensions. The goal of SOMs is to convert N-dimensional data as something that can be visually understood more easily (Figure 6.16).

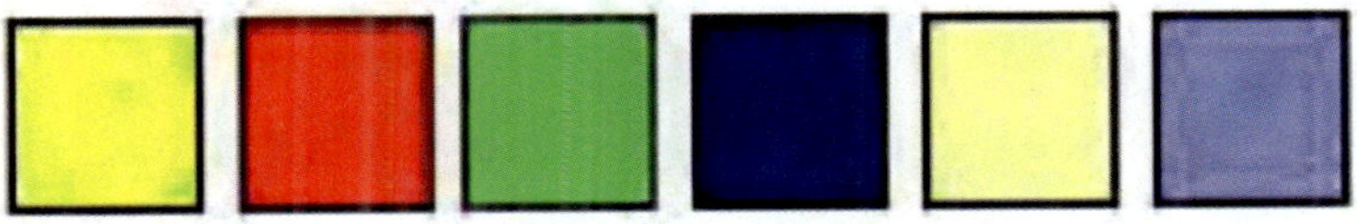

Figure 6.16. Illustration of sample data via the self-organized map.

Source: https://www.intechopen.com/books/new-advances-in-machine-learning/types-of-machine-learning-algorithms.

As a result, one would anticipate that the pink, blue, and gray colors would end up close to each other on the map, and the yellow color would end up close to both the green and red colors. The weight vector is the second component of the Self-Organize Maps model. Essentially, every vector is made up of two components, which are illustrated in the following explanation. Among the weight vector's components are two components that are related to the data: the first is the data, and the second is its natural location. The most advantageous aspect of color is that it allows for the presentation of weight vector data through the use of color. A pixel's (x, y) position on the screen reflects the data in this case, and its color denotes the location.

To illustrate this, a 2D array of weight vectors was employed, which would look similar to the image shown in Figure 6.16. This image depicts a slanted perspective of the grid in which the n-dimensional array for each weight takes precedence, and each weight vector has its own exclusive location in the grid, as seen in the previous image. The weight vectors do not have to be established in two dimensions in order to be effective. A significant amount of work has already been completed with the help of self-organized maps (SOMs) with only a single dimension. When compared to the sample vectors, the data fragment of the weight vector must have the same dimensions as the sample vectors. Due to the fact that the SOMs are basically neural networks, the weights are sometimes referred to as neurons in some contexts. For the demonstration of examples, the method in which the SOMs assemble themselves is frequently a difficult challenge to understand. In addition, the neurons are given the ability to alter themselves by adapting to develop more like the samples in the hopes of winning the upcoming competition. This learning and selection mechanism causes the weights to form themselves into a map that represents resemblances between the two objects being studied (Figure 6.17).

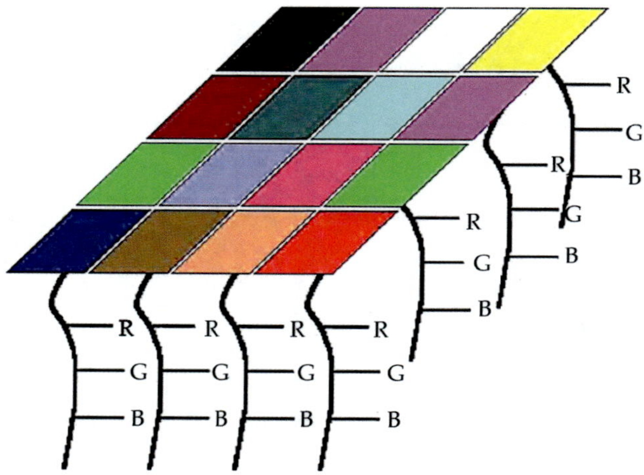

Figure 6.17. Depiction of 2D array weight of a vector.

Source: https://www.intechopen.com/books/new-advances-in-machine-learning/types-of-machine-learning-algorithms.

So, how are the weight vectors organized to indicate the resemblances of sample vectors with the sample and weight vectors? This is accomplished by following the basic procedure outlined in Figure 6.18.

```
Initialize Map
For t from 0 to 1

Randomly select a sample
Get best matching unit
Scale neighbors
Increase t a small amount

End for
```

Figure 6.18. Illustration of a sample SOM algorithm.

Source: https://www.intechopen.com/books/new-advances-in-machine-learning/types-of-machine-learning-algorithms.

It is necessary to initialize the weight vectors in order to begin creating the self-organized map, and this is the first stage in the process. Following initialization, a sample vector may be chosen at random, and then the map of weight vectors can be searched in order to discover the weight that most closely fits the selected sample vector. Along with its specific position, each weight vector contains a set of surrounding weights that are close to it. The chosen weight is balanced by the fact that it has the potential to become increasingly similar to the arbitrarily picked sample vector. The neighbors of a certain weight are additionally compensated by the fact that they are capable of becoming increasingly similar to the sample vector that was chosen. After this step, t is raised by a small amount due to the fact that the number of neighbors and the amount to which each weight can acquire reduces with time t is increased by a tiny amount after this step. After then, the entire operation is done approximately 1,000 times more. In the condition of colors, the processes involved in implementing the simple algorithm are given in subsections.

6.4.5.2. Initializing of the Weights

The weight vector map is initialized in three ways, as seen in the pictures below. The java application below shows the six different blue, red, and green intensities. Because the true values for the weights are frequently floating, the weights have a larger array than the six values shown in the diagram (Figure 6.19).

Figure 6.19. Demonstration of weight values.

Source: https://www.intechopen.com/books/new-advances-in-machine-learning/types-of-machine-learning-algorithms.

A variety of different approaches are available for initializing the weight vector. The first way is to assign arbitrary values to each weight vector based on the data it contains. The screen of pixels with arbitrary blue, green, and red values is shown on the left side of the screen, as seen above. Calculating the SOMs as per Kohonen is a computationally-intensive process; as a result, different options to initialize the weight exist in order to ensure that the samples are not identical. This approach is time-saving because it requires fewer iterations to produce a usable map, and it is less time-consuming.

In addition to the arbitrary technique, two alternative ways for initializing the weight are being researched and developed. To get this effect, one approach is to place different colors at each of the four corners and then allow them to fade gently towards the center. One technique includes placing the colors blue, red, and green at similar distances from one another and also away from the center of the image.

6.4.5.3. Attainment of Best Matching Unit

Performing this step is straightforward; simply go through all of the weights and compute the distance between the selected sample vector and each weight. The weight with the least gap between it and the others is the winner. The winning weight vector is chosen at random from among the weight vectors that have the smallest distance if more than one weight vectors have the same difference in distance from each other. There are a variety of different approaches that may be used to define what distance means mathematically. Euclidean distance is the most usual method used:

$$\sqrt{\sum_{i=0}^{n} x_i^2}$$

$x[i] \rightarrow$ data value at i^{th} data fellow of the sample.

$n \rightarrow$ number of dimensions for the sample vectors.

Colors may be thought of as 3D points, with each portion acting as an axis. Now, if we choose the green color (0, 6, 0), the light green color (3, 6, 3) will be considerably closer to our chosen green color than red with value (0, 6, 0).

Light Green = Sqrt((3–0)^2+(6–6)^2+(3–0)^2) = 4.24

Red = Sqrt((6–0)^2+(0–6)^2+(0–0)^2) = 8.49

As a result, the light green hue is the best feasible match. This procedure of computing and associating distances is carried out over the whole map. The winner is the weight vector with the least distance. The square root is not computed in the java application for performance optimization of this part.

6.4.5.4. Scale Neighbors

When it comes to the scaling of nearby weights, there are two basic components to consider. It is necessary to determine which of the weights are considered to be neighbors, as well as to what degree each weight vector can become comparable to the sample vector, in order to complete the first portion. Different strategies may be used to determine the neighbors of the winning weight vector, and each method has its own advantages and disadvantages. While some approaches make use of concentric squares, other ways make use of hexagons. One of the ways that make use of the Gaussian function is presented, in which any single point with a value greater than zero is considered to be a neighbor.

Throughout this chapter, it has been mentioned that the number of neighbors is decreasing with time. This is done in order for the samples to be able to go to the zone where they are most likely to be found and then compete for the position. Similar to course alteration followed by fine-tuning, this method is carried out in the same way. It doesn't really matter what function is used to lower the radius of the effect, as long as it reduces the radius of the effect (Figure 6.20).

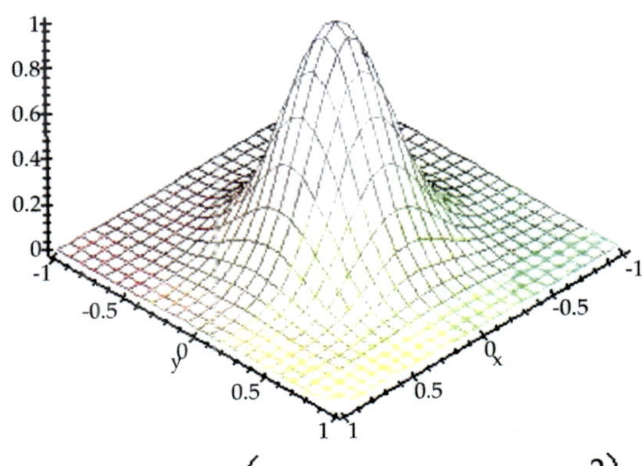

$$f := (x,y) \rightarrow e^{\left(-6.666666667 \sqrt{x^2+y^2}^{\,2}\right)}$$

Figure 6.20. A graph demonstrating the determination of SOM neighbor.

Source: https://www.intechopen.com/books/new-advances-in-machine-learning/types-of-machine-learning-algorithms.

The plot of the function utilized is shown in the diagram above. With the passage of time, the base moves closer to the center, resulting in fewer neighbors as time goes on. The initial radius is set to a number that is near the map's width or height.

The second aspect of the scaling of neighbors is the learning function. By becoming comparable to the sample vector, the winning weight vector is compensated. The neighbors begin to resemble the sample vector as well. This learning system has the property that the further a neighbor is from winning weight, the less it will learn. The percentage at which the weight may learn decreases, and it can be adjusted to any desirable range. The Gaussian function employed returns a value between 0 and 1, which is then changed using the parametric equation for each neighbor. After then, the new hue will be given a name:

*Current color**(1.–*t*) + *sample vector***t*

As the number of weight vector neighbors decreases, the amount of information that the weight vector can learn decreases as well with the passage of time, as shown in Figure 6.20. The winning weight vector is converted into the sample vector on the very first iteration because t has the full range from 0 to 1 on the very first iteration. As time goes on, the winning weight vector begins to resemble the sample vector to a greater or lesser extent, with the highest value of t diminishing in importance as well. The rate at which the weight vector can learn decreases linearly as the number of weight vectors increases. To demonstrate this, consider the preceding plot, where the amount of information a weight vector can learn is equal to the height of the bump. With the passing of time, the height of the bump will gradually decrease in height.

The neighbors of a particular weight vector are identified once it has been determined that it is the winner. Every individual neighbor, and also the winning weight vectors, undergoes a transformation to become more similar to the sample vector.

6.4.5.5. Determination of SOMs' Quality

Below is another example of a self-organized map generated by the program after 500 iterations. You'll see in the diagram that the colors that are similar are grouped together once more. This isn't always the case, as some colors are surrounded by colors that aren't similar to them. Because we are familiar with colors, it is simple to draw attention to them. When more abstract data is used, determining whether two entities are similar because they are close to each other becomes extremely difficult (Figure 6.21).

Figure 6.21. Display of SOM iterations.

Source: https://www.intechopen.com/books/new-advances-in-machine-learning/types-of-machine-learning-algorithms.

There is a fairly easy way for demonstrating where the similarities and differences exist. To calculate this, we must first go over all of the weights and then determine how similar the neighbors are to one another. The distance traveled by the weight vectors between each weight and its neighbors is then computed. The color is assigned to that specific site using an average of the computed distances. This process is placed in Screen.java with the name of *public void update_bw()*.

If the determined average distance is quite great, the weights near to one other are different, and a dark hue is assigned to the weight location at that particular place. For those instances where the calculated average distance is little, the lighter color is assigned. Because the color of the blobs' centers is the same as the surrounding region, it must be white because the same color is present in the surrounding area. It shouldn't be white in the sections between the blobs when there are commonalities between the blobs. It has to be a light gray hue to be effective. When blobs are considerably next to one other but are not identical, a black color must be used to separate them (Figure 6.22).

Figure 6.22. A sample of weight allocation in colors.

Source: https://www.intechopen.com/books/new-advances-in-machine-learning/types-of-machine-learning-algorithms.

The gorges of the black color display where the color may be considered close to one another on the map in the image shown above. When we're talking about weights with real values, the colors aren't all the same. The areas between the blobs where a light gray color is present represent an exact match. The black color is existent in the right bottom of the image, which is surrounded by colors that are not comparable to it. When looking at the black and white Self-organized map, it's clear that the black color isn't comparable to the other. There is no relation between the two colors represented by the black lines. The pink color is present in the topmost corner, with a light green color adjacent to it. Both colors aren't close to each other in reality, but the colored Self-organized map depicts them as such.

We can essentially assign a value to each Self-organized map based on the average distances used to create black and white SOMs, which will determine how well the image has depicted the likenesses of the samples.

6.4.5.6. Pros and Cons of SOM (Self-Organized Map)

Some of the pros of SOMs are:

- SOMs are self-explanatory, which is one of the finest features of the Self-organized map. SOMs are pretty comparable if they are quite close to one other and a gray connects them. SOMs are drastically different if there is a dark ravine between them. People can quickly learn how to utilize SOMS in an effective manner, unlike Multidimensional Scaling.

- Another feature of SOMs is their ability to work quickly and efficiently. SOMs are highly good at classifying data and are easy to evaluate in terms of their overall quality. The efficiency of a map may be calculated, and the similarities between items can be computed using this method.
- Some of the cons of SOMs are:
- One of the most difficult aspects of creating SOMs is obtaining accurate data. Unfortunately, in order to create a map, you must have a value for each and every dimension of each and every member of the sample collection. Sometimes this is just not possible, and more often than not, obtaining the information is a time-consuming endeavor. This is the limiting property of self-organized maps, and it is referred to as the missing data problem in the literature.
- Unlike the commonalities between the sample vectors, each of the SOMs is unique and learns something new. SOMs are used to organize sample data so that when the final output is completed, the samples are generally surrounded by samples that are comparable to each other. Similar samples, on the other hand, are not always closely related. The creation of a vast number of maps is required to generate a single good map.
- The third issue with SOMs is the cost of calculation, which is the primary drawback. As data dimensions get larger, dimension reduction visualization approaches become more important, and the time it takes to calculate the dimensions grows as well.

REFERENCES

1. Abraham, T. H., (2004). Nicolas Rashevsky's mathematical biophysics. *Journal of the History of Biology, 37*(2), 333–385.
2. Acuna, E., & Rodriguez, C., (2004). The treatment of missing values and its effect on classifier accuracy. In: *Classification, Clustering, and Data Mining Applications* (Vol. 1, pp. 639–647). Springer, Berlin, Heidelberg.
3. Ahangi, A., Karamnejad, M., Mohammadi, N., Ebrahimpour, R., & Bagheri, N., (2013). Multiple classifier system for EEG signal classification with application to brain–computer interfaces. *Neural Computing and Applications, 23*(5), 1319–1327.
4. Aldape-Pérez, M., Yáñez-Márquez, C., Camacho-Nieto, O., López-Yáñez, I., & Argüelles-Cruz, A. J., (2015). Collaborative learning based on associative models: Application to pattern classification in medical datasets. *Computers in Human Behavior, 51*, 771–779.
5. Allix, N. M., (2000). The theory of multiple intelligences: A case of missing cognitive matter. *Australian Journal of Education, 44*(3), 272–288.
6. Allix, N. M., (2003). Epistemology and knowledge management concepts and practices. *Journal of Knowledge Management Practice, 1*, 1–20.
7. Alpaydin, E., (2004). *Introduction to Machine Learning.* Massachusetts, USA: MIT Press.
8. Alpaydm, E., (1999). Combined 5× 2 cv F test for comparing supervised classification learning algorithms. *Neural Computation, 11*(8), 1885–1892.
9. Ambikairajah, E., Keane, M., Kelly, A., Kilmartin, L., & Tattersall, G., (1993). Predictive models for speaker verification. *Speech communication, 13*(3, 4), 417–425.
10. Anevski, D., Gill, R. D., & Zohren, S., (2013). *Estimating a Probability Mass Function with Unknown Labels.* arXiv preprint arXiv:1312.1200.
11. Anil, M. G. P., (1999). Socialization influences on preparation for later life. *Journal of Marketing Practice: Applied Marketing Science, 5*(6, 7, 8), 163–176.
12. Anuradha, C., & Velmurugan, T., (2014). A data mining based survey on student performance evaluation system. In: *Computational Intelligence*

and Computing Research (ICCIC), 2014 IEEE International Conference (pp. 1–4). IEEE.

13. Ashby, W. R., (1960). *Design of a Brain, The Origin of Adaptive Behavior.* John Wiley and Son.
14. Ayodele, T. O., (2010). Machine learning overview. In: *New Advances in Machine Learning*. InTech.
15. Bader-El-Den, M., & Poli, R., (2007). Generating SAT local-search heuristics using a GP hyper-heuristic framework. In: *International Conference on Artificial Evolution (Evolution Artificielle)* (pp. 37–49). Springer, Berlin, Heidelberg.
16. Bateson, G., (1960). Minimal requirements for a theory of schizophrenia. *AMA Archives of General Psychiatry, 2*(5), 477–491.
17. Batista, G., (2003). An analysis of four missing data treatment methods for supervised learning. *Applied Artificial Intelligence, 17*, 519–533.
18. Bilmes, J. A., (1998). A gentle tutorial of the EM algorithm and its application to parameter estimation for gaussian mixture and hidden Markov models. *International Computer Science Institute, 4*(510), 126.
19. Bishop, C. M., (1995). *Neural Networks for Pattern Recognition.* Oxford, England: Oxford University Press.
20. Bishop, C. M., (2006). *Pattern Recognition and Machine Learning (Information Science and Statistics).* New York, New York: Springer Science and Business Media.
21. Block, H. D., (1961). *The Perceptron: A Model of Brian Functioning, 34*(1), 123–135.
22. Block, H. D., Knight, Jr. B. W., & Rosenblatt, F., (1962). Analysis of a four-layer series-coupled perceptron. II. *Reviews of Modern Physics, 34*(1), 135.
23. Brereton, R. G., (2015). Pattern recognition in chemometrics. *Chemometrics and Intelligent Laboratory Systems, 149*, 90–96.
24. Burke, E. K., Hyde, M. R., & Kendall, G., (2006). Evolving bin packing heuristics with genetic programming. In: *Parallel Problem Solving from Nature-PPSN IX* (pp. 860–869). Springer, Berlin, Heidelberg.
25. Campbell, D. T., (1976). On the conflicts between biological and social evolution and between psychology and moral tradition. *Zygon®, 11*(3), 167–208.

26. Carling, A., (1992). *Introducing Neural Networks.* Wilmslow, UK: Sigma Press.
27. Caudill, M., & Butler, C., (1993). *Understanding Neural Networks: Computer Explorations (No. 006.3 C3) (Vol.1, pp.1-25).*
28. Chali, Y., & Joty, S. R., (2009). Complex question answering: Unsupervised learning approaches and experiments. *Journal of Artificial Intelligent Research,* 1–47.
29. Chen, Y. S., Qin, Y. S., Xiang, Y. G., Zhong, J. X., & Jiao, X. L., (2011). Intrusion detection system based on immune algorithm and support vector machine in wireless sensor network. In: *Information and Automation* (pp. 372–376). Springer, Berlin, Heidelberg.
30. D. Michie, D. J., (1994). *Machine Learning, Neural and Statistical Classification.* Prentice Hall Inc.
31. Dempster, A. P., Laird, N. M., & Rubin, D. B., (1977). Maximum likelihood from incomplete data via the EM algorithm. *Journal of the Royal Statistical Society; Series B (Methodological)*, 1–38.
32. Dockens, W. S., (1979). Induction/catastrophe theory: A behavioral ecological approach to cognition in human individuals. *Systems Research and Behavioral Science, 24*(2), 94–111.
33. Doya, K., (2007). *Bayesian Brain: Probabilistic Approaches to Neural Coding*. MIT press.
34. Durbin, R., & Rumelhart, D. E., (1989). Product units: A computationally powerful and biologically plausible extension to backpropagation networks. *Neural Computation, 1*(1), 133–142.
35. Dutra Da, S. R., Robson, W., & Pedrini, S. H., (2011). Image segmentation based on wavelet feature descriptor and dimensionality reduction applied to remote sensing. *Chilean J. Stat., 2.*
36. Dutton, J. M., & Starbuck, W. H., (1971). Computer simulation models of human behavior: A history of an intellectual technology. *IEEE Transactions on Systems, Man, and Cybernetics*, (2), 128–171.
37. Elliott, S. W., & Anderson, J. R., (1995). *Learning and Memory.* Wiley, New York, USA.
38. Farhangfar, A., Kurgan, L., & Dy, J., (2008). Impact of imputation of missing values on classification error for discrete data. *Pattern Recognition, 41*(12), 3692–3705.
39. Fausett, L., (1994). *Fundamentals of Neural Networks.* New York: Prentice Hall.

40. Fisher, D. H., (1987). Knowledge acquisition via incremental conceptual clustering. *Machine Learning, 2*(2), 139–172.
41. Forsyth, M. E., Hochberg, M., Cook, G., Renals, S., Robinson, T., Schechtman, R., & Doherty-Sneddon, G., (1995). Semi-continuous hidden {M}arkov models for speaker verification. In: *Proc. ARPA Spoken Language Technology Workshop* (Vol. 1, pp. 2171–2174). Universiteit Twente, Enschede.
42. Forsyth, R. S., (1990). The strange story of the perceptron. *Artificial Intelligence Review, 4*(2), 147–155.
43. Franc, V., & Hlaváč, V., (2005). Simple solvers for large quadratic programming tasks. In: *Joint Pattern Recognition Symposium* (pp. 75–84). Springer, Berlin, Heidelberg.
44. Freund, Y., & Schapire, R. E., (1999). Large margin classification using the perceptron algorithm. *Machine Learning, 37*(3), 277–296.
45. Friedberg, R. M., (1958). A learning machine: Part, 1. *IBM Journal,* 2–13.
46. Fu, S. S., & Lee, M. K., (2005). IT based knowledge sharing and organizational trust: The development and initial test of a comprehensive model. *ECIS 2005 Proceedings,* 56.
47. Fukunaga, A. S., (2008). Automated discovery of local search heuristics for satisfiability testing. *Evolutionary Computation, 16*(1), 31–61.
48. Gennari, J. H., Langley, P., & Fisher, D., (1988). *Models of Incremental Concept Formation (No. UCI-ICS-TR-88-16).* California Univ. Irvine Dept. of Information and Computer Science.
49. Getoor, L., & Taskar, B., (2007). *Introduction to Statistical Relational Learning*. MIT press.
50. Ghahramani, Z., (2008). *Unsupervised Learning Algorithms are Designed to Extract Structure from Data* (Vol. 178, pp. 1–8). IOS Press.
51. Ghahramani, Z., (2008). *Unsupervised Learning Algorithms are Designed to Extract Structure from Data*, 178.
52. Gillies, D., (1996). *Artificial Intelligence and Scientific Method.* OUP Oxford.
53. Gong, Y., & Haton, J. P., (1992). Non linear vectorial interpolation for speaker recognition. In: *Proceedings of IEEE Int. Conf. Acoust., Speech, and Signal Processing* (Vol. 2, pp. II173–II176). San Francisco, California, USA.

54. González, L., Angulo, C., Velasco, F., & Catala, A., (2006). Dual unification of bi-class support vector machine formulations. *Pattern Recognition, 39*(7), 1325–1332.
55. Goodacre, R., Broadhurst, D., Smilde, A. K., Kristal, B. S., Baker, J. D., Beger, R., & Ebbels, T., (2007). Proposed minimum reporting standards for data analysis in metabolomics. *Metabolomics, 3*(3), 231–241.
56. Gregan-Paxton, J., Hoeffler, S., & Zhao, M., (2005). When categorization is ambiguous: Factors that facilitate the use of a multiple category inference strategy. *Journal of Consumer Psychology, 15*(2), 127–140.
57. Gross, G. N., Lømo, T., & Sveen, O., (1969). *Participation of Inhibitory and Excitatory Interneurones in the Control of Hippocampal Cortical Output*. Per Anderson, The Interneuron.
58. Grzymala-Busse, J. W., & Hu, M., (2000). A comparison of several approaches to missing attribute values in data mining. In: *International Conference on Rough Sets and Current Trends in Computing* (pp. 378–385). Springer, Berlin, Heidelberg.
59. Grzymala-Busse, J. W., Goodwin, L. K., Grzymala-Busse, W. J., & Zheng, X., (2005). Handling missing attribute values in preterm birth data sets. In: *International Workshop on Rough Sets, Fuzzy Sets, Data Mining, and Granular-Soft Computing* (pp. 342–351). Springer, Berlin, Heidelberg.
60. Hall, M., Frank, E., Holmes, G., Pfahringer, B., Reutemann, P., & Witten, I. H., (2009). The WEKA data mining software: An update. *ACM SIGKDD Explorations Newsletter, 11*(1), 10–18.
61. Hammes, M., & Wieland, R., (2012). Screening tool to stress while working. In: Athanassiou, G., Schreiber-Costa, S., & Sträter, O., (eds.), *Psychology of Occupational Safety and Health-Successful Design of Safe and Good Work-Research and Implementation in Practice* (pp. 331–334). Kröning: Asanger.
62. Hastie, T., Tibshirani, R., & Friedman, J., (2009). Unsupervised learning. In: *The Elements of Statistical Learning* (pp. 485–585). Springer, New York, NY.
63. Hattori, H., (1992). Text independent speaker recognition using neural networks. In: *Proceedings of IEEE Int. Conf. Acoust., Speech, and Signal Processing* (Vol. 2, pp. II153–II156). San Francisco, California, USA.

64. Haykin, S., & Network, N., (2004). A comprehensive foundation. *Neural Networks, 2*(2004), 41.
65. Haykin, S., (1994). *Neural Networks: A Comprehensive Foundation.* New York: Macmillan Publishing.
66. Herlihy, B., (1998). Targeting 50+ mining the wealth of an established generation. *Direct Marketing, 61*(7), 18–20.
67. Higgins, A., Bahler, L., & Porter, J (1996). Voice identification using nonparametric density matching. In: Lee, C. H., Soong, F. K., & Paliwal, K. K., (eds.), *Automatic Speech and Speaker Recognition* (pp. 211–233.
68. Hodge, V. A., (2004). A survey of outlier detection methodologies. *Artificial Intelligence Review, 22*(2), 85–126.
69. Holland, J., (1980). *Adaptive Algorithms for Discovering and Using General Patterns in Growing Knowledge Bases Policy Analysis and Information Systems, 4*(3).
70. Honghai, F., Guoshun, C., Cheng, Y., Bingru, Y., & Yumei, C., (2005). A SVM regression based approach to filling in missing values. In: *International Conference on Knowledge-Based and Intelligent Information and Engineering Systems* (pp. 581–587). Springer, Berlin, Heidelberg.
71. Hopfield, J. J., (1982). Neural networks and physical systems with emergent collective computational abilities. *Proceedings of the National Academy of Sciences, 79*(8), 2554–2558.
72. Hornik, K., Stinchcombe, M., & White, H., (1989). Multilayer feedforward networks are universal approximators. *Neural Networks, 2*(5), 359–366.
73. Hunt, E. B., (1966). *Experiment in Induction.* (pp.1-10)
74. Ian, H. W., & Eibe. F., (2005). *Data Mining Practical Machine Learning and Techniques* (2nd edn.). Morgan Kaufmann.
75. Jaime, G. C., & Michalski R. S., (1983). Machine learning: A historical and methodological analysis. *Association for the Advancement of Artificial Intelligence, 4*(3), 1–10.
76. Jayaraman, P. P., Zaslavsky, A., & Delsing, J., (2010). Intelligent processing of k-nearest neighbors queries using mobile data collectors in a location aware 3D wireless sensor network. In: *International Conference on Industrial, Engineering and Other Applications of Applied Intelligent Systems* (pp. 260–270). Springer, Berlin, Heidelberg.

77. Kandjani, H., Bernus, P., & Nielsen, S., (2013). Enterprise architecture cybernetics and the edge of chaos: Sustaining enterprises as complex systems in complex business environments. In: *System Sciences (HICSS), 2013 46th Hawaii International Conference* (pp. 3858–3867). IEEE.
78. Keller, R. E., & Poli, R., (2007). Linear genetic programming of parsimonious metaheuristics. In: *Evolutionary Computation, 2007, CEC 2007; IEEE Congress* (pp. 4508–4515). IEEE.
79. Kim, M. H., & Park, M. G., (2009). Bayesian statistical modeling of system energy saving effectiveness for MAC protocols of wireless sensor networks. In: *Software Engineering, Artificial Intelligence, Networking and Parallel/Distributed Computing* (pp. 233–245). Springer, Berlin, Heidelberg.
80. Knill, D. C., & Richards, W., (1996). *Perception as Bayesian Inference*. Cambridge University Press.
81. Kohonen, T., (1997). *Self-Organizating Maps*. (Vol.1, pp.1-25)
82. Konopka, A. K., (2006). *Systems Biology: Principles, Methods, and Concepts*. CRC Press.
83. Lakshminarayan, K., Harp, S. A., & Samad, T., (1999). Imputation of missing data in industrial databases. *Applied Intelligence, 11*(3), 259–275.
84. Lebowitz, M., (1987). Experiments with incremental concept formation: UNIMEM. *Machine Learning, 2*(2), 103–138.
85. LeCun, Y., Boser, B., Denker, J. S., Henderson, D., Howard, R. E., Hubbard, W., & Jackel, L. D., (1989). Backpropagation applied to handwritten zip code recognition. *Neural Computation, 1*(4), 541–551.
86. Li, D., Deogun, J., Spaulding, W., & Shuart, B., (2004). Towards missing data imputation: A study of fuzzy k-means clustering method. In: *International Conference on Rough Sets and Current Trends in Computing* (pp. 573–579). Springer, Berlin, Heidelberg.
87. Long, G. E., (1980). Surface approximation: A deterministic approach to modelling spatially variable systems. *Ecological Modeling, 8*, 333–343.
88. López, J. E. N., Castro, G. M., Saez, P. L., & Muiña, F. E. G., (2002). An integrated model of creation and transformation of knowledge [CD-ROM] In: *Memory of the XXII Symposium on the Management of*

Technological Innovation of the Nucleus of Policy and Technological Management. University of São Paulo.

89. López, P. V., (2001). The information society in Latin America and the Caribbean: ICTs and a new institutional framework [CD-ROM]. *Report of the 9th Latin-Ibero-American Seminar on Technological Management Innovation in the Knowledge Economy.*
90. Luengo, J., García, S., & Herrera, F., (2012). On the choice of the best imputation methods for missing values considering three groups of classification methods. *Knowledge and Information Systems, 32*(1), 77–108.
91. Luis, G. L. A., (2005). Unified dual for bi-class SVM approaches. *Pattern Recognition, 38*(10), 1772–1774.
92. Lukasiak, B. M., Zomer, S., Brereton, R. G., Faria, R., & Duncan, J. C., (2007). Pattern recognition and feature selection for the discrimination between grades of commercial plastics. *Chemometrics and Intelligent Laboratory Systems, 87*(1), 18–25.
93. Lula, P., (2000). *Selected Applications of Artificial Neural Networks using Statistica Neural Networks*. StatSoft, Krakow, Poland.
94. Mamassian, P., Landy, M., & Maloney, L. T., (2002). Bayesian modeling of visual perception. *Probabilistic Models of the Brain,* 13–36.
95. McCulloch, W. S., (1943). A logical calculus of the ideas immanent in nervous activity. *Bull. Math. Biophysics,* 115–133.
96. Michalski, R. S., & Stepp, R., (1982). *Revealing Conceptual Structure in Data by Inductive Inference (Vol.1, pp.1-20).*
97. Michalski, R. S., & Stepp, R. E., (1983). *Learning from Observation: Conceptual Clustering.* TIOGA Publishing Co.
98. Minsky, M., & Papert, S. A., (2017). *Perceptrons: An Introduction to Computational Geometry.* MIT Press.
99. Mitchell, T. M., (2006). *The Discipline of Machine Learning.* Machine Learning Deparment technical report CMU-ML-06-108, Carnegie Mellon University.
100. Mitra, S., Datta, S., Perkins, T., & Michailidis, G., (2008). *Introduction to Machine Learning and Bioinformatics*. CRC Press.
101. Mohammad, T. Z., & Mahmoud, A. M., (2014). Clustering of slow learners behavior for discovery of optimal patterns of learning. *Literatures, 5*(11).

102. Mooney, R. J., (2000). Learning language in logic. In: Science, L. N., (ed.), *Learning for Semantic Interpretation: Scaling Up without Dumbing Down* (pp. 219–234). Springer Berlin/Heidelberg.
103. Mooney, R. J., (2000). Learning language in logic. *LN Science, Learning for Semantic Interpretation: Scaling Up without Dumbing Down,* 219–234.
104. Mostow, D., (1983). *Transforming Declarative Advice into Effective Procedures: A Heuristic Search cxamplc In I?. S. Michalski.* Tioga Press.
105. Mostow, D., (1983). Transforming declarative advice into effective procedures: A heuristic search cxamplc In I?
106. Neymark, Y., Batalova, Z., & Obraztsova, N., (1970). Pattern recognition and computer systems. *Engineering Cybernetics, 8*, 97.
107. Nilsson, N. J., (1982). *Principles of Artificial Intelligence (Symbolic Computation/Artificial Intelligence).* Springer.
108. Noguchi, S., & Nagasawa, K., (2014). New concepts, that is, the information processing capacity and the a% information processing capacity are introduced. *Methodologies of Pattern Recognition,* 437.
109. Novikoff, A. B., (1963). *On Convergence Proofs for Perceptrons.* Stanford research Inst Menlo Park CA.
110. Olga, R., Deng, J., Su, H., Krause, J., Satheesh, S., Ma, S., & Berg, A. C., (2015). *Large Scale Visual Recognition Challenge.* ImageNet http://arxiv. org/abs/1409.0575.
111. Oltean, M., & Dumitrescu, D., (2004). Evolving TSP heuristics using multi expression programming. In: *International Conference on Computational Science* (pp. 670–673). Springer, Berlin, Heidelberg.
112. Oltean, M., & Groşan, C., (2003). Evolving evolutionary algorithms using multi expression programming. In: *European Conference on Artificial Life* (pp. 651–658). Springer, Berlin, Heidelberg.
113. Oltean, M., (2005). *Evolving Evolutionary Algorithms Using Linear Genetic Programming, 13*(3), 387 –410.
114. Oltean, M., (2007). Evolving evolutionary algorithms with patterns. *Soft Computing, 11*(6), 503–518.
115. Orlitsky, A., Santhanam, N., Viswanathan, K., & Zhang, J., (2005). Convergence of profile-based estimators. *Proceedings of International Symposium on Information Theory. Proceedings. International Symposium* (pp. 1843–1847). Adelaide, Australia: IEEE.

116. Orlitsky, A., Santhanam, N., Viswanathan, K., & Zhang, J., (2006). Theoretical and experimental results on modeling low probabilities. In: *Information Theory Workshop, 2006, ITW'06 Punta del Este; IEEE* (pp. 242–246). IEEE.

117. Otair, M. A., & Salameh, W. A., (2004). An improved back-propagation neural networks using a modified nonlinear function. In: *Proceedings of the IASTED International Conference* (pp. 442–447).

118. Patterson, D., (1996). *Artificial Neural Networks.* Singapore: Prentice Hall.

119. Pickering, A., (2002). Cybernetics and the mangle: Ashby, beer and pask. *Social Studies of Science, 32*(3), 413–437.

120. Poli, R., Woodward, J., & Burke, E. K., (2007). A histogram-matching approach to the evolution of bin-packing strategies. In: *Evolutionary Computation, 2007, CEC 2007; IEEE Congress* (pp. 3500–3507). IEEE.

121. Pollack, J. B., (1989). Connectionism: Past, present, and future. *Artificial Intelligence Review, 3*(1), 3–20.

122. Punitha, S. C., Thangaiah, P. R. J., & Punithavalli, M., (2014). Performance analysis of clustering using partitioning and hierarchical clustering techniques. *International Journal of Database Theory and Application, 7*(6), 233–240.

123. Purves, D., & Lotto, R. B., (2003). *Why We See What We Do: An Empirical Theory of Vision*. Sinauer Associates.

124. Puterman, M. L., (2014). *Markov Decision Processes: Discrete Stochastic Dynamic Programming*. John Wiley & Sons.

125. Rajesh, P. N. R., (2002). *Probabilistic Models of the Brain.* MIT Press.

126. Rao, R. P., (2005). Bayesian inference and attentional modulation in the visual cortex. *Neuroreport, 16*(16), 1843–1848.

127. Rashevsky, N., (1948). *Mathematical Biophysics: Physico-Mathematical Foundations of Biology.* Chicago: Univ. of Chicago Press.

128. Rebentrost, P., Mohseni, M., & Lloyd, S., (2014). Quantum support vector machine for big data classification. *Physical Review Letters, 113*(13), 130503.

129. Richard, O. D., & David P (2000). *Pattern Classification* (2nd edn.).

130. Richard, S. S., & Barto, A. G., (1998). *Reinforcement Learning.* MIT Press.
131. Ripley, B., (1996). *Pattern Recognition and Neural Networks.* Cambridge University Press.
132. Rosenblatt, F., (1958). The perceptron: A probabilistic model for information storage and organization in the brain. *Psychological Review, 65*(6), 386–408.
133. Rosenblatt, F., (1961). *Principles of Neurodynamics Unclassifie—Armed Services Technical Information Agency*. Spartan, Washington, DC.
134. Rumelhart, D. E., Hinton, G. E., & Williams, R. J., (1985). *Learning Internal Representations by Error Propagation (No. ICS-8506).* California Univ San Diego La Jolla Inst for Cognitive Science.
135. Rumelhart, D. E., Hinton, G. E., & Williams, R. J., (1986). Learning representations by back-propagating errors. *Nature, 323*(6088), 533.
136. Russell, S. J., & Norvig, P., (2016). *Artificial Intelligence: A Modern Approach*. Malaysia; Pearson Education Limited.
137. Russell, S. J., (2003). *Artificial Intelligence: A Modern Approach* (2nd edn.). Upper Saddle River, NJ, NJ, USA: Prentice Hall.
138. Ryszard, S. M. (1955). *Machine Learning: An Artificial Intelligence Approach* (Vol. I). Morgan Kaufmann.
139. Sakamoto, Y., Jones, M., & Love, B. C., (2008). Putting the psychology back into psychological models: Mechanistic versus rational approaches. *Memory & Cognition, 36*(6), 1057–1065.
140. Sanchez, R., (1997). Strategic management at the point of inflection: Systems, complexity and competence theory. *Long Range Planning, 30*(6), 939–946.
141. Schaffalitzky, F., & Zisserman, A., (2004). Automated scene matching in movies. CIVR 2004. *Proceedings of the Challenge of Image and Video Retrieval.* London, LNCS, 2383.
142. Schwenker, F., & Trentin, E., (2014). Pattern classification and clustering: A review of partially supervised learning approaches. *Pattern Recognition Letters, 37*, 4–14.
143. Seising, R., & Tabacchi, M. E., (2013). A very brief history of soft computing: Fuzzy sets, artificial neural networks and evolutionary computation. In: *IFSA World Congress and NAFIPS Annual Meeting (IFSA/NAFIPS), 2013 Joint* (pp. 739–744). IEEE.

144. Selfridge, O. G., (1959). Pandemonium: A paradigm for learning. In: *The Mechanization of Thought Processes*. H.M.S.O., London. London.

145. Shmailov, M. M., (2016). Breaking through the iron curtain. In: *Intellectual Pursuits of Nicolas Rashevsky* (pp. 93–131). Birkhäuser, Cham.

146. Shmailov, M. M., (2016). Scientific experiment: Attempts to converse across disciplinary boundaries using the method of approximation. In: *Intellectual Pursuits of Nicolas Rashevsky* (pp. 65–92). Birkhäuser, Cham.

147. Singh, S., & Lal, S. P., (2013). Educational courseware evaluation using machine learning techniques. In: *E-Learning, e-Management and e-Services (IC3e), 2013 IEEE Conference* (pp. 73–78). IEEE.

148. Sleeman, D. H., (1983). *Inferring Student Models for Intelligent CAI; Machine Learning.* Tioga Press.

149. Somorjai, R. L., Dolenko, B., & Baumgartner, R., (2003). Class prediction and discovery using gene microarray and proteomics mass spectroscopy data: Curses, caveats, cautions. *Bioinformatics, 19*(12), 1484–1491.

150. Stepp, R. E., & Michalski, R. S., (1986). *Conceptual Clustering: Inventing Goal-Oriented Classifications of Structured Objects. (Vol.1,pp.1-25)*

151. Stewart, N., & Brown, G. D., (2004). Sequence effects in the categorization of tones varying in frequency. *Journal of Experimental Psychology: Learning, Memory, and Cognition, 30*(2), 416.

152. Sutton, R. S., & Barto, A. G., (1998). *Introduction to Reinforcement Learning* (Vol. 135). Cambridge: MIT press.

153. Sutton, R. S., (1988). Learning to predict by the methods of temporal differences. *Machine Learning, 3*(1), 9–44.

154. Tapas, K. D. M., (2002). A local search approximation algorithm for k-means clustering. *Proceedings of the Eighteenth Annual Symposium on Computational Geometry* (pp. 10–18). Barcelona, Spain: ACM Press.

155. Tavares, J., Machado, P., Cardoso, A., Pereira, F. B., & Costa, E., (2004). On the evolution of evolutionary algorithms. In: *European Conference on Genetic Programming* (pp. 389–398). Springer, Berlin, Heidelberg.

156. Teather, L. A., (2006). Pathophysiological effects of inflammatory mediators and stress on distinct memory systems. In: *Nutrients, Stress, and Medical Disorders* (pp. 377–386). Humana Press.
157. Timothy, J. S. P. J., (1998). Decision fusion using a multi-linear classifier. In: *Proceedings of the International Conference on Multisource-Multisensor Information Fusion*.
158. Tom, M., (1997). *Machine Learning*. Machine Learning, Tom Mitchell, McGraw Hill, 1997: McGraw Hill.
159. Trevor, H. R. T., (2001). *The Elements of Statistical Learning*. New York, NY, USA: Springer Science and Business Media.
160. Turnbull, S., (2002). The science of corporate governance. *Corporate Governance: An International Review, 10*(4), 261–277.
161. Vancouver, J. B., (1996). Living systems theory as a paradigm for organizational behavior: Understanding humans, organizations, and social processes. *Systems Research and Behavioral Science, 41*(3), 165–204.
162. Wagh, S. P., (1994). *Intonation Knowledge Based Speaker Recognition Using Neural Networks*. MTech. project report, Indian Institute of Technology, Department of Electrical Engineering.
163. Weiss, Y., & Fleet, D. J., (2002). Velocity likelihoods in biological and machine vision. *Probabilistic Models of the Brain: Perception and Neural Function,* 81–100.
164. Widrow, B. W., (2007). *Adaptive Inverse Control: A Signal Processing Approach*. Wiley-IEEE Press.
165. Xu, L., & Jordan, M. I., (1996). On convergence properties of the EM algorithm for Gaussian mixtures. *Neural Computation, 8*(1), 129–151.
166. Yang, Z., & Purves, D., (2004). The statistical structure of natural light patterns determines perceived light intensity. *Proceedings of the National Academy of Sciences of the United States of America, 101*(23), 8745–8750.
167. Yu, L. L., (2004). Efficient feature selection via analysis of relevance and redundancy. *JMLR*, 1205–1224.
168. Yusupov, T., (2007). *The Efficient Market Hypothesis Through the Eyes of an Artificial Technical Analyst*. Doctoral dissertation, Christian-Albrechts Universität Kiel.

169. Zanibbi, R., & Blostein, D., (2012). Recognition and retrieval of mathematical expressions. *International Journal on Document Analysis and Recognition (IJDAR), 15*(4), 331–357.

170. Zeiler, M. D., & Fergus, R (2013). Stochastic pooling for regularization of deep convolutional. *Neural Netw. Google Scholar.*

171. Zhang, S. Z., (2002). Data preparation for data mining. *Applied Artificial Intelligence, 17*, 375–381.

172. Zoltan-Csaba, M., Pangercic, D., Blodow, N., & Beetz, M., (2011). Combined 2D-3D categorization and classification for multimodal perception systems. *Int. J. Robotics Res. Arch, 30*(11).

CHAPTER 7

APPROXIMATION ALGORITHMS

CONTENTS

7.1. INTRODUCTION

From the computational perspective, the main fascinating daily optimization tasks have been incredibly hard. Usually, finding a near-optimum or optimum solution to a larger-scale optimization of the issue requires extra computer sources that are available. There is a substantial body of literature that examines how the computational strains of a solution technique rise to have the issue size to be expressed to uncover the computational qualities of different optimization problems (Alon and Spencer, 2000). A difference is drawn between issues that need computing resources that expand polynomial with the size of the problem and those whereby the resources required rise exponentially. Issues in the first kind are called effectively resolvable; while, issues in the second category are referred to as intractable as the exponential rise in essential computer sources renders although the smaller examples of these issues non-solvable (Cook and Rohe, 1999; Chazelle et al., 2001; Carlson et al., 2006).

The classifications of a substantial number of popular optimization problems as NP-hard are included. It's generally not thought yet confirmed (Clay Mathematics Institute, 2003) those NP-hard problems are inflexible, implying that no efficient algorithm (such as one that measures polynomials) can be guaranteed to provide optimum solutions. The minimal bin-packing problem, the smallest traveling salesman issue, and the minimal graph coloring problem are instances of NP-hard optimization tasks. Because of the nature of NP-hard problems, an advancement that guides to a greater understanding of their computational characteristics, makeup, and methods for resolving anyone, approximately or perfectly, results in enhanced algorithms for many NP-hard issues. Several computing problems are discovered to be NP-hard in fields as diverse as computer-aided design and finance, economics, biology, and operations research.

An obvious question is whether near-optimal solutions to challenging optimization problems like these may be attained efficiently. Simulated annealing and tabu research are 2 heuristic local research approaches that are generally extremely successful in finding approximate solutions. Such strategies, on the other hand, don't approach with any assurances about the class of the absolute solution or the needed utmost running time. We'll talk about the most theoretical techniques to such problems employing referred to as "approximation algorithms," which are effective algorithms that may be proved to yield high-quality answers. We would also address problem types for which no effective approximation algorithms are available, showing a significant portion of the discussion for the widely used heuristic local search approaches (Shaw et al., 1998; Dotu et al., 2003).

The structure of reasonable algorithms of approximation is an extremely vigorous area of search in which novel techniques and approaches are being discovered regularly. Such strategies can become more important in the context of solving large-scale optimization challenges encountered daily (Feller, 1971; Hochbaum, 1996; Cormen et al., 2001).

When bin packing and multiprocessor scheduling were first introduced in the last 1960s and the early 1970s, a specific concept of approximation had been planned in the domain of bin-packing (Graham, 1966; Garey et al., 1972, 1976; Johnson, 1974). Usually, 2 characteristics are shared by all approximation algorithms. In the first instance, they give a suitable solution to a difficult scenario in a polynomial amount of time. The development of a technique that identifies some sort of logical answer is not difficult in the majority of cases (Kozen, 1992). The second feature of approximation algorithms describes how we are concerned with having a certain class of solutions that are guaranteed. The class of an approximation method is defined as the "distance" among the ideal solutions and its solutions, measured across all of the potential causes of the issue, which is the greatest possible distance. An algorithm approximates the resolution of an optimization issue if it gives a plausible solution with a measurement that is closer to optimum consistently, for instance within a variable restricted by a fixed or within a progressively rising function of the input size. For a given minimization problem Π, under the assumption of constant α, an algorithm A is considered an α-approximation algorithm if its response is as close to the optimum as feasible, taking into account all probable sources of problem Π.

We will focus on the drawing of approximation techniques for the NP-hard optimization issue. We'll illustrate how common algorithm drawing methods like greedy approaches and local research are applied to create fine approximation algorithms. We'll demonstrate how randomization may be used to devise approximation techniques. Randomized algorithms are exciting as they are generally simpler to implement, examine, and implement quicker as compared to deterministic algorithms (Motwani and Raghavan, 1995). A random algorithm makes several of its decisions randomly; it "flips a coin" to decide on what to act at certain phases. Even though looking at a similar scenario of a problem, numerous implementations of a randomized algorithm might result in various solutions and running times due to its random element. We'll show how to use randomization and approximation algorithms to effectively estimated NP-hard optimization issues. The approximation algorithm's running time, the approximation ratio and the approximation solution may all be random variables in this situation. When faced with an optimization issue, the goal is to construct a randomized approximation technique that is provably bounded by a polynomial whose potential answer is, in probability, close to the perfect solution. These guarantees pertain to every situation of the problem being controlled. The only arbitrariness in the randomized approximation algorithm's efficiency guaranty comes from the algorithm itself, not from the examples.

Because there are no effective algorithms for discovering optimal solutions for NP-hard issues, a key topic is whether we may compute exceptional approximations which are closer to ideal. It will be extremely attractive (and realistic) if one might go from exponential to polynomial-time difficulty through loosening the check on optimality, particularly if the error is kept to a minimum (Figure 7.1) (Qi, 1988; Vangheluwe et al., 2003; Xu, 2005).

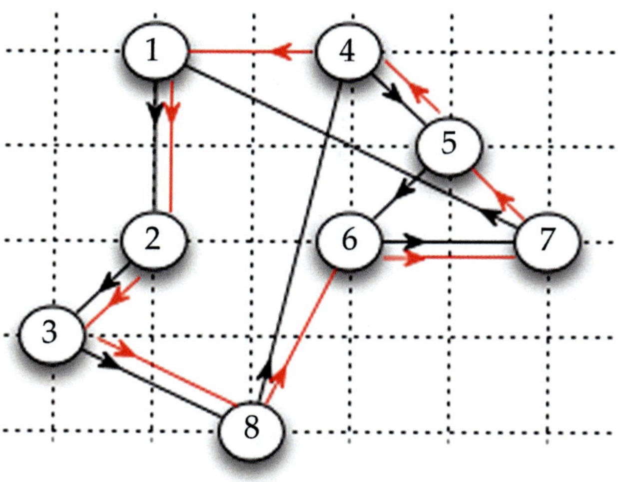

Figure 7.1. The mechanism for approximation algorithms shown schematically.

Source: http://faculty.ycp.edu/~dbabcock/PastCourses/cs360/lectures/lecture29.html.

It has been argued that good approximation algorithms may be used to solve some of the most difficult issues in combinational optimization. So, the APX intricacy category consists of problems that may be approximated by a polynomial-time approximation algorithm having the ratio of performance that is constrained by a constant. We may devise even the best approximation algorithms for certain issues if we prepare ahead of time. More precisely, we can consider a collection of approximation algorithms that allow us to come as close to the optimal as we desire while still being willing to exchange quality for time (Reinelt, 1994; Indrani, 2003). A polynomial-time approximation scheme (PTAS) is a specific class of algorithms that gages polynomials in the degree of input. The PTAS classification in the category of optimization problems that may be solved on behalf of an approximation scheme (AS) that gages polynomials in the degree of input. Occasionally, we may design approximation systems that gage polynomial, both in terms of their accuracy and in the words of the size of the input data they receive. Specifically, we are referring to the class of problems that are amenable to this form of completely polynomial-time approximation strategy by using FPTAS (Boyd and Pulleyblank, 1990; Gomes and Shmoys, 2002).

Furthermore, the approximations that are obtained so far for certain NP-hard problems are insufficient, and in other cases, no one can devise approximation methods in the optimal constant factor (Hochbaum and Shmoys, 1987). It wasn't clear at first whether the poor results were due to our inability to come up with appropriate approximation methods for this sort of problem or to a certain inherent structural feature of the issues that prevent them with reasonable approximations. We'll observe that an approximation has restrictions that are intrinsic to certain types of issues. For instance, in some cases, the approximation constant factor has a lower constraint, while in others; we may presumably represent that there are no estimations under a certain constant factor of the optimal. In essence, there is a broad spectrum of cases ranging from NP-hard optimization issues that permit approximations to some extent to issues that don't allow approximations in any way. We'll give a quick rundown of the proof strategies utilized to get non-approximately conclusions (Figure 7.2) (Pulleyblank, 1989).

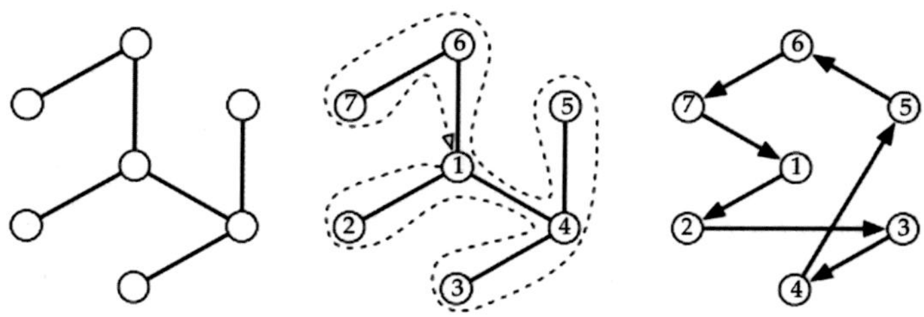

Figure 7.2. Approximation route for an approximation algorithm problem.

Source: http://fliphtml5.com/czsc/chmn/basic.

Using instances of algorithms with such qualities, we feel that the best method to comprehend the concepts of randomization and approximation is to explore cases of algorithms with such properties. We would introduce the intuitive notion in each section, and then emphasize its prominent elements with carefully chosen examples of archetypal situations in the following segment (Podsakoff et al., 2009; Bennett et al., 2015). We are not attempting to give a full analysis of algorithms of the approximation or the optimal algorithms of the approximation for the situations that we have presented here in this chapter. Various evaluation and design approaches for randomized and approximation algorithms are instead described, with evident examples that allow for straightforward explanations. It is possible to obtain approximations with improved performance guarantees for several problems; however, this needs more complicated proving methods that are outside of the scope of this beginning tutorial. In these kinds of scenarios, we would direct the reader to the related literature outputs to learn more (Spyke, 1998; Becchetti et al., 2006).

7.2. APPROXIMATION STRATEGIES

7.2.1. Optimization Problems

We shall traditionally discuss optimization problems (Ausiello et al., 1999). The requirement of a feasible solution to the issue, the setup of the input example, and the evaluation function utilized to select that viable solutions are decided to be optimal are the 3 distinguishing aspects of any optimization problem. The problem title would reveal if we want a viable solution having a minimal or maximal measure. The minimal vertex cover issue can be characterized in the following way (Leahu and Gomes, 2004).

- Minimum Vertex Cover:

Case: A graph that is undirected G = (V, E).

Solution: A subset S ⊆ V so that for each {u,v} ∈ E, either u ∈ S or v ∈ S.

Measure: |S|.

For things connected to Case I, we employ the following system:

- The series of viable solutions to I is Sol(I);
- m_I: Sol(I) → R is the calculate function concomitant having I;
- Opt(I) ⊆ Sol(I) is the viable solution having optimal calculation (either minimal or maximal).

As a result, we can fully indicate an optimization problem Π by giving a set of tuples {(I, Sol(I), m_I, Opt(I))} that encompasses all feasible cases I. It's vital to realize that Sol (I) and I might refer to fully distinct domains. The group of I represents all meaningless graphs, whereas Sol (I) represents all conceivable subdivisions of vertices in a graph (Vershynin, 2009).

7.2.2. Approximation and Performance

Simply put, an algorithm approximates the resolution of an optimization issue if it makes a viable solution that is closer to optimum in measure at all times. Below, this intuition is proven correct. Assume Π a problem of optimization. We suppose that when a case I ∈ Π, A(I) ∈ Sol(I); i.e., is described, an algorithm (A) plausibly solves Π, and A delivers a potential solution to us.

Allow A to feasibly resolve. The approximation ratio α (A) of A is defined as the smallest feasible ratio among the measurement of A (I) and the extent of an optimum solution. Formally,

$$\alpha(A) = \min_{I \in \Pi} \frac{m_I(A(I))}{m_I(Opt(I))}$$

For minimization issues, this ratio is always at least 1. It is always at extreme 1 for maximization issues.

7.2.3. Complex Background

A decision problem is described as an optimization issue with a measure of 0 to 1. Such that, responding to a *no* or *yes* question regarding I is part of addressing a case I of a decision problem (where no correlates to a measure of 0, and yes correlates to a measure of 1). As a result, a decision problem can be thought of as a subgroup S of the larger group of all conceivable cases, with members of S denoting examples of value 1.

On any "rational" model of computation, P (polynomial time) is viewed as the category of decision problems Π with a persistent algorithm A_{Π} whereby each example I ∈ Π is sorted by A_{Π} in a polynomial (|I|k for certain constant k) the total number of steps. Pointer machines, random access machines, single-tape, and multi-tape Turing machines are examples of rational models (Belanger and Wang, 1993).

Whereas P refers to a category of issues Π that may be resolved quickly, NP (non-deterministic polynomial-time) refers to a category of decision problems Π that may be effectively evaluated. NP, on the other hand, in the category of decision issues with a persistent decision problem Π^0 in P and a constant k satisfying:

I ∈Π if and only if there exists C ∈ $\{0, 1\}^{|I|k}$ like (I, C) ∈Π'

In another sense, if a case I am in is an NP issue, one may deduce whether a specific short string C of polynomial length in me may be successfully solved. Assume the NP problem of finding if a graph G contains a route P that passes via all nodes precisely once (this is known as the Hamiltonian path issue) (Johnson, 1973; Ho, 1982; Blass and Gurevich, 1990). If given G and an explanation of P, it is relatively simple to confirm that P is such a pathway by ensuring that:

- P consists of all nodes in G;
- No node shows greater than one in G;
- Every 2 nearby nodes in P have a competitive advantage in G.

The fundamental distinction between NP and P is that it is not recognized how to discover this type of path P given only a graph G. In fact, the Hamiltonian route issue exists not just in NP, but also in NP-hard.

It's worth noting that, whereas a short proof is always present if I ∈ Π, it can't be the case that short proofs are present for examples that aren't in Π. As a result, P issues are those that are effectively decidable, whereas NP problems are those that are effectively verifiable by a short proof (Hopper and Chazelle, 2004).

We'll also consider NPO and PO, which are the optimization equivalents of NP and P, correspondingly. Informally, PO is a group of optimization problems with a polynomial-time algorithm that always produces an optimum solution in each case of the problem, while NPO is a group of optimization problems wherein the measurement function is polynomial-time quantifiable, and an algorithm may calculate whether or not a probable solution is possible in polynomial time.

We'll concentrate on approximation solutions to the "toughest" NPO situations, those with an NP-hard consistent decision problem. There are some NPO issues of this type that may be estimated extremely precisely, while others may scarcely be approximated at all (Jiménez et al., 2001; Cueto et al., 2003; Aistleitner, 2011).

7.3. THE GREEDY METHOD

Greedy approximation algorithms are intended with a straightforward philosophy in mind: make decisions over and over again that get one closer to a viable solution to the issue. According to a deficient although simply quantifiable heuristic, such options will be the best ones to choose from. The short-term advantage of this heuristic is that it tries to be as opportunistic as is reasonably conceivable. Therefore, these algorithms are referred to as greedy; yet, a more appropriate phrase may be "short-sighted." Consider the following scenario: the goal is to discover the fastest route from my house to the theater (Klein and Young, 2010; Ausiello et al., 2012). If I think that the walkthrough Forbes Avenue is about a similar distance as the walkthrough 5th Avenue, and I am closer to Forbes Avenue than 5th Avenue, it will make sense to travel towards Forbes Avenue and use that pathway rather than Fifth Avenue.

The success of this plan is dependent on my opinion that the Forbes way is just as honorable as the 5th path, which I believe to be right. We would demonstrate that, for certain issues, selecting a solution according to opportunistic, imprecise heuristic results in a non-trivial approximation procedure that is difficult to implement (Karp, 1975; Paz and Moram, 1977; Mossel et al., 2005).

7.3.1. Greedy Vertex Cover

The minimal vertex cover issue was discussed in detail throughout the preliminary rounds. In diverse fields of optimization study, several solutions to the problem are proposed. To solve the problem, we would develop a straightforward greedy approach that is a 2-approximation; that is, the vertex cover cardinality resumed through our algorithm is no more than double the cardinality of the lowest cover (Khot, 2002; Khot et al., 2007; Khot and Vishnoni, 2015). The following is the algorithm.

Greedy-VC: To begin, consider S to be a empty set, select a random edge between {u,v}. S is formed by adding u and v to S and removing u and v from the graph. Repeat the process until no edges are visible in the graph. Yield S as the vertex cover in this case.

- Theorem 1: Greedy-VC is a 2-Approximation Algorithm for Minimum Vertex Cover

Proof: To begin, we assert that S, as resumed by Greedy-VC, is a vertex cover. Assume that this is not the case; then there is an edge e that had not protected by any vertex in S. Because we only eliminate vertices in S from the graph, an edge e will remain in the graph after Greedy-VC is concluded, that is a contradiction (Liu, 1976; Durand et al., 2005).

Allow S^* to be the smallest vertex cover. We'll now say that S^* has at least $|S|/2$ vertices. Because it would trail $|S^*| \geq |S|/2$, that's whys our algorithm uses a $|S|/|S^*| \leq 2$ approximation ratio.

Because the edges we select in Greedy-VC don't share any endpoints, we may conclude that:

- $S|/2$ denotes the total edges quantity we select;
- S^* should have to choose a minimum of one vertex from every edge we select.

As a result, we get $|S^*| \geq |S|/2$.

When verifying that an algorithm has a defined approximation ratio, the evaluation may be rather "loose," and cannot reflect the best probable ratio which may be attained. It demonstrates that Greedy-VC is not superior to a 2-approximation. Particularly, in the case of complete bipartite graphs, there is an endless set of Vertex Cover situations where Greedy-VC provably chooses precisely double the number of vertices essential to cover the graph (Book and Siekmann, 1986; Hermann and Pichler, 2008).

On Vertex Cover, one final point should be made. Even though the aforementioned approach is very basic, no better approximation algorithms exist! It is commonly recognized that unless P = NP, minimum vertex cover may not be approximated best than $2 - \varepsilon$ for any $\varepsilon > 0$ (Figure 7.3) (Hermann and Kolaitis, 1994; Khot and Regev, 2003).

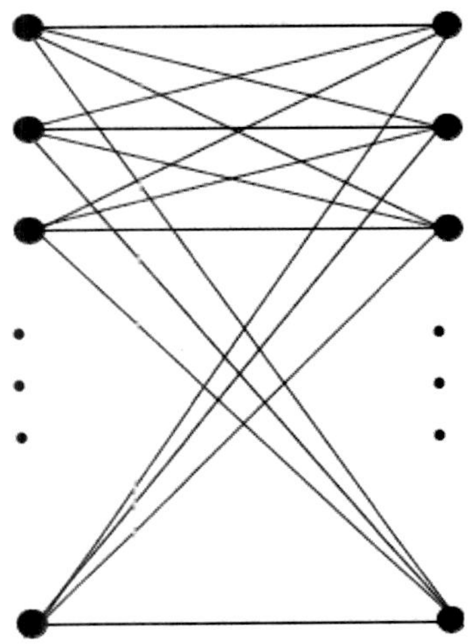

Figure 7.3. A diagram of a completed bipartite graph with n nodes colored red and n nodes colored blue.

Source: https://link.springer.com/chapter/10.1007/0-387-28356-0_18.

A graph for which its vertices can be allotted one of two colors is termed as *bipartite* (say, *blue* or *red*), in such a manner that all edges have different colored endpoints. When applying Greedy-VC on such cases (for any normal number n), the algorithm would pick all 2_n vertices.

7.3.2. Greedy MAX-SAT

MAX-SAT is a well-studied issue with versions appearing in a variety of discrete optimization fields. It needs some terminology, to begin with. We'll only work with Boolean variables (those that are either false or true), which we'll express with x_1, x_2, and so on. A literal is referred to as the negation of a variable or the variable itself (for example, x_7, x_{11} are literals). A sentence is referred to as the OR of a few literals (for example, $(-x_1 \vee x_7 \vee x_{11})$). If a Boolean formula is provided as an AND of clauses, we presume it is in CNF (conjunctive normal form). For example, $(-x_1 \vee x_7 \vee -x_{11}) \wedge (x_5 \vee -x_2 \vee -x_3)$ is in CNF. Ultimately, the MAX-SAT issue entails assigning a value to the variables of the Boolean formula in CNF as the largest number of clauses are either fixed to true or satisfied. Accurately:

- **MAX-SAT:** A Boolean formula F in CNF is an example.

Solution: An assignment a, that is a function from every variable in F to false or true.

The number of clauses in F which are set to true (satisfied) when the variables in F are allocated according to a.

What is a typical greedy technique for approximate MAXSAT resolution? One strategy is to select a variable that, when assigned to a defined value, fulfills numerous clauses. If a variable is invalid in numerous clauses, it is natural to assume that setting the variable to *false* would satisfy all of them; so, this technique should essentially fix the problem. Let $n(l_i, F)$ be the total number of sentences in F that include the literal l_i.

- **Greedy-MAXSAT:** Select a literal l_i with the highest $n(l_i, F)$ value. Set the corresponding factor of l_i to true to satisfy all sentences with l_i, resulting in a lower F. Rep until none of the variables remains in F.

Greedy-MAXSAT runs in polynomial time, which is straightforward to understand (coarsely quadratic time, contingent with the computational model picked for analysis). For the MAX-SAT issue, it's also a "decent" approximation.

7.3.3. Greedy MAX-CUT

The next instance shows how approximation algorithms can be designed using local research (particularly, hill climbing). Hill climbing is inherently a greedy endeavor: given a potential solution x, one strives to enhance it via selecting a possible y that is "near" to x but provides a better measure (higher or lower, based upon maximization or minimization). Constant attempts at enhancement usually result in "locally" optimum solutions with a decent measure when compared to a globally optimum solution (that is a member of Opt (I)). We describe local search by introducing an approximation technique that finds a locally optimum significant assignment for the NP-complete MAX-CUT problem. It's crucial to realize that not all local research procedures aim to locate a local optimum simulated annealing, for instance, aims to avoid local optimum in the hopes of discoveries a worldwide optimum (Ghalil, 1974; Černý, 1985).

- MAX-CUT:

Case: A non-directed graph $G = (V,E)$.

Solution: A graph cut, which is, a pair (S, T) such that $S \subseteq V$ and $T = V - S$.

Measure: The *size of the cut*, which is the edges quantity intersecting the cut, such that $|\{\{u,v\} \in E \mid u \in S, v \in T\}|$.

Our local research technique enhances the existing potential answer by varying the location of one vertex in the cut until no more enhancements are available. We'll show that for such a local maximal, the size of the cut is minimum m/2.

- **Local-Cut:** Begin with a V cut at random. Determine whether moving every vertex to the extra side of the division enhances the cut size. If this is the case, modify. Repeat until there are no more likely motions.

To begin, keep in mind that this method repeats at most m times, because every movement of a vertex enhances the size of the cut by at least 1, and a cat may be as large as m.

- **Theorem 2: Local-Cut is an a$\frac{1}{2}$-Approximation Algorithm for MAX-CUT:**

Proof: Assume (S, T) is the algorithm's cut, and assume v as a vertex. After the algorithm completion, check to see if the total number of continuous edges that cross (S, T) is larger as compared to the total number of contiguous edges which does not cross; or else, v will have been shifted. Assume that deg(v) is the level of (v). Our observation shows that deg(v)/2, on the other hand, restricts the algorithm's cut off of v cross.

Suppose m be the number of intersecting edges in the returned cut. Because every crossing edge has 2 endpoints, the sum/counts every intersecting edge at most twice such as:

$$\sum_{v \in V} (\deg(v)/2) \leq 2m^*$$

Although, note $\sum_{v \in V} \deg(v) = 2m;$ When all the degrees of vertices are added up, every edge is recorded twice, once for each terminus. As a result, we may say:

$$m = \sum_{v \in V} (\deg(v)/2) \leq 2m^*$$

The algorithm has the following approximation ratio $\frac{m^*}{m} \geq \frac{1}{2}$.

MAX-CUT appears to grant significantly best approximation ratios than 1/2; a claimed relaxing of the problem to a semi-certain linear program yields a 0.8786 approximation ratio (Goemans and Williamson, 1995).

Furthermore, MAX-CUT, such as numerous other optimization problems ($1 - \varepsilon$, for all $\varepsilon > 0$) except P = NP, may not be approximated arbitrarily. To put it another way, it's doubtful that MAX-CUT appears in the PTAS complication class.

7.3.4. Greedy Knapsack

In operations research, the knapsack issue and its variants have been comprehensively studied. The concept is simple: you have a C-capacity knapsack and a set of objects 1, …, n. Every object has a cost c_i associated with transporting it, as well as a profit p_i associated with carrying it. The goal is to locate a subset of items having a cost of no more than C while maximizing profit (Edmonds, 1965; Halperin, 2002).

- Maximum Integer Knapsack:

Case: A capability $C \in N$, and various objects $n \in N$, with persistent profits and costs c_i, $p_i \in N$ for all $i = 1, \ldots, n$.

Solution: A subset $S \subseteq \{1, \ldots, n\}$ so that $Pj_{\in S}\ c_j \leq C$.

Measure: The *overall profit* $\sum j_{\in S}\ p_j$.

Maximum Integer Knapsack is NP-hard. This issue may also be resolved in polynomial time in a "fractional" version (we called it maximum fraction knapsack). Instead of having to select the whole product, this form allows you to select items fractions, such as 1/8 of the 1st item, 1/2 of the 2nd item, and so on. The cost and profit of the objects would be fractional as well (1/8 of the profit and cost of the first, 1/2 of the cost and profit of the 2nd, and so on) (Miller, 1976; Geman and Geman, 1987).

One greedy plan for resolving these 2 issues is to box items having the biggest profit-to-cost ratio first, hoping to get several smaller cost and higher profit objects in the knapsack. It turns out that such an algorithm would not provide any constant approximation guarantee; rather a tiny variant on this strategy will provide a 2-approximation for Integer Knapsack, and a precise algorithm for Fraction Knapsack (Adleman, 1980; Lenstra et al., 1990). The algorithms for Integer Knapsack and Fraction Knapsack are, respectively:

- **Greedy-IKS:** Pick items having the biggest profit-to-cost ratio 1st, till the overall cost of items picked is larger as compared to C. Suppose j be the final object is selected, and S be the group of items picked before j. Return either {j} or S, contingent on which one is more beneficial.

- **Greedy-FKS:** Pick items as in Greedy-IKS. While item j marks the cost of the existing solution larger as compared to C, improve the *fraction* of j so that the resultant solution cost is precisely C.

We skip the proof of the succeeding. An occupied treatment may be shown in (Ausiello et al., 1999).

7.4. SEQUENTIAL ALGORITHMS

For approximations on the issue, whereas a viable solution is a scenario dividing into subsets, sequential methods are used. A sequential approach "sorts" the case's objects in a certain way, and then divides the case into partitions depending upon such ordering (Zhu and Wilhelm, 2006; Wang, 2008).

7.4.1. Sequential Bin Packing

We start with a problem called Minimum Packaging of Bin, which is related to the knapsack problem.

- Minimum Bin Packing:

Case: A series of objects $S = \{r_1, \ldots, r_n\}$, where $r_i \in (0,1]$ for all i = 1, …, n.

Solution: Splitting of S into bins $B_1, \ldots, B_M$ so that $\sum rj_{\in B}i\ r_j \leq 1$ for all i = 1, …, M. *Measure*: M.

A well-known algorithm for minimum packaging of bins remains an online technique. Allow for j = 1 at first, and a bin B_1 to be accessible. Pack the novel object ri into the previous bin used, B_j, as one goes over the input (r_1, r_2, etc.). If r_i isn't right for B_j, construct a new bin, $B_j + 1$, and place a_i in it. This method is "online" because it works on the input in a consistent sequence, therefore adding additional things to the case while its running does not affect the output (Stock and Watson, 2001).

- Theorem 3: Last-Bin is a 2-Approximation to Minimum Bin Packing:

Proof: Assume that m is the maximum bins quantity utilized by the method and that m is the smallest possible bins quantity for the current situation. Notice that $m^* \geq R$, since the overall bins quantity necessary is as a minimum equal to the total mass of all objects (each bin embraces 1 unit). Furthermore, given any pair of bins, B_i and B_{i+1}, generated by the algorithm, the total number of items from 'S' in B_i and $B_i + 1$ must equal at least 1; otherwise, we will have retained the items from B_{i+1} in B_i. This demonstrates that $m \leq 2R$. As a result, $m \leq 2R \leq 2m^*$, and the algorithm is a two-approximation (Berry and Howls, 2012).

Building a set of instances demonstrating how this approximation bound, identical to the one for Greedy-VC, is built and is an exciting exercise for the reader. Some algorithms produce the best approximations. While trying to pack an a_i, for instance, we don't even consider the preceding bins $B_1, \ldots, B_{j-1}$; just the latest one is taken into account (Arora et al., 2001).

Consider the following change to Last-Bin as a result of this concept, choose every a_i article in lessening size order and place it in the *1st accessible* bin from $B_1, \ldots, B_j$. (If ai can't suit in any of the prior j bins, a novel bin is merely formed.) This new algorithm is known as First-Bin. A comprehensive approximation of situations may be used to create a better approximation bound.

7.4.2. Sequential Job Scheduling

One of the most important issues in the theory of scheduling is determining how to distribute work among several machines in a way that all of the jobs are completed effectively. In this case, we'll look at completing the project in the smallest quantity of time feasible. For the sake of abstraction and ease, we'll assume that the machines distribute power in the same way for every job.

- Minimum Job Scheduling:

Case: An integer k and a multi-set $T = \{t_1, \ldots, t_n\}$ of *times*, $t_i \in Q$ for all $i = 1, \ldots, n$ (*which is,* the t_i are fractions).

Solution: A function ranging from $\{1, \ldots, n\}$ to $\{1, \ldots, k\}$ that allocates jobs to machines.

Measure: The overall time it takes for all machines to complete a task if they execute in parallel: max $\{\sum i_{:a(i)=j}\ t_i \mid j \in \{1, \ldots, k\}\}$.

When reading a novel task through time t_i, allocate it to the machine j that recently has the lowest aggregation of job, that is, the j with minimal Pi: $_{an}(_i) = _j t_i$. Sequential-Jobs are the name of the algorithm.

- Theorem 4: Sequential Tasks be a 2-Approximation Meant for Minimum Job Scheduling:

Proof: Suppose j be a machine with the fastest time of completion and let "*i*" be the catalog of the algorithm's most recent task assignment to j. Let $s_{i,j}$ be the total number of times all tasks preceding "*i*" are allocated to

j. (This can be taken to represent the moment when job "*i*" on machine j starts.) Because the method allocated "*i*" to the machine with the minimum amount of work, all other machines "j" currently has greater $\sum i: {}_{an}(_{i}) = {}_{j}0\ t_{i}$. As a result, $s_{i,j} \leq \frac{1}{k}\sum^{n}_{i=1} t_{i}$, implying that $s_{i,j}$ is less than 1/k of the overall time spent on entire tasks (remember k is the total number of machines).

Because the sum applies to the scenario when each machine takes precisely a similar proportion of time to finish, $B = \frac{1}{k} \sum^{n}_{i=1} t_{i} \leq m^{*}$ is the achievement time for an ideal solution. Thus, the accomplishment time for machine *j* is:

$s_{i,j} + t_{i} \leq m* + m* = 2m*$

As a result, the maximum completion time is double as long as the best solution. This isn't the best option: Minimum Job Scheduling has PTAS as well (Vazirani, 1983).

7.5. RANDOMIZATION

For algorithmic design, randomness is a valuable resource. A broad range of unique mathematics can be used to help the examination of an algorithm on the supposition that one has accessibility to impartial coins that can be flipped and their values (tails or heads) extracted. It's not uncommon for a basic randomized algorithm to provide similar performance guarantees as a sophisticated deterministic (such as non-randomized) method.

Adding randomness to a computer process can occasionally result in a significant speedup over strictly deterministic procedures, according to one of the majority exciting findings in the field of algorithm planning. The following observations can be used to intuitively express this. On a collection of deterministic algorithms, a randomized algorithm may be seen as a probability allotment. A randomized algorithm's behavior might vary based upon the random selections completed by the algorithm; as a result, when we think of a randomized algorithm, we're thinking of a randomly selected algorithm from a collection of algorithms. If a significant fraction of these deterministic algorithms carries out well on the provided input, continuing the randomized method after a certain point in runtime would result in a speedup (Gomes et al., 1998).

Certain random algorithms, such as polynomial identity checking, can effectively resolve issues for which no efficient deterministic method exists (Motwani and Raghavan, 1995). In the widely used simulated annealing approach for resolving optimization issues, randomization is also critical

(Kirkpatrick et al., 1983). Finally, the challenge of determining whether a given number is prime (a crucial problem in modern cryptography) could only be solved effectively through randomization. A deterministic method primarily was recently revealed (Agrawal et al., 2002).

7.5.1. Random MAX-CUT Solution

A random technique for MAXCUT that provides a 2-approximation was previously shown. We may propose an exceedingly tiny approximation technique that is a similar performance in approximation and runs in anticipated polynomial time by utilizing randomization.

Random-Cut: Select a cut at random (such as random division of the vertices into 2 groups). Repeat if lesser than m/2 edges are intersecting this cut.

For MAX-CUT, Random-Cut remains a $\frac{1}{2}$-approximation method that runs in polynomial time.

Proof: Assume X is a random factor that represents the edges quantity that intersects a cut. X_i is a pointer factor which is one when the ith edge intersects the cut and zero or else for i = 1, …, m. Afterwards $X = \sum_{i=1}^{m} X_i$, so by the linearity of m probability. $E[X] = \sum_{i=1}^{m} E[X_i]$.

Now, each edge {u, v} has a 1/2 probability of intersecting arbitrarily selected cut. (Why? We put u and v in one of 2 probable partitions at random, so u and v would be in the same partition with probability 1/2.) As a result, $E[X_i] = 1/2$ for every "i" resulting in E[X] = m/2.

This just demonstrates that if we choose a random cut, we may anticipate minimum m/2 edges to cross. We needed a randomized method that always produces a decent cut and whose running duration is an arbitrary variable with a polynomial expectation. Let's see how likely it is that $X \geq m/2$ will be found once a random cut is made. When $X \geq m/2$ is the worst case, and it's all probability is dependent upon m, and when $X < m/2$ is the best case, all likelihood is dependent upon $m/2-1$. This raises the probability of getting an at least-m/2 cut while lowering the probability of getting X. Formally,

$$m/2 = E[X] \leq (1 - \Pr[X \geq m/2])(m/2 - 1) + \Pr[X \geq m/2]m$$

When solving for $\Pr[X \geq m/2]$, the least value is 2/ (m + 2). As a result, the predictable number of repeats in the above approach is at most (m + 2)/2; as a result, the procedure runs in polynomial time and always produces a cut of size minimum m/2.

We see that if we stated our approximation as "select a random stop and cut," the method would run in linear time and have an expected approximation ratio of 1/2.

7.5.2. Random MAX-SAT Solution

We studied a random MAX-SAT technique that had been guaranteed to satisfy half of the clauses. We'll look at MAX-Ak-SAT, which restricts MAX-SAT to CNF principles with a minimum of k literals per clause. Our method is quite identical to MAXCUT's: Select a random assignment for the variables. Utilizing an equivalent approach to the above notion, it is simple to show that the expected approximation ratio of such a method is at least $1 - 1/2^k$. If m is the number of clauses in a formulary, then $m - m/2^k$ is the anticipated number of clauses satisfied by an arbitrary assignment.

Let c be a clause with k literals that are chosen randomly. Because every literal has a probability of 1/2 and there is a minimum of k of them, the probability that each of them was fixed to a number that stamps them false is as high as $1/2^k$. As a result, the probability of c being pleased is at least $1-1/2^k$. We find that $m - m/2^k$ clauses are predicted to be met by utilizing the linearity of the probability argument (as in the MAX-CUT analysis).

REFERENCES

1. Adleman, L. M., (1980). On distinguishing prime numbers from composite numbers. In: *Foundations of Computer Science, 1980, 21st Annual Symposium* (pp. 387–406). IEEE.
2. Agrawal, M., Kayal, N., & Saxena, N., (2002). PRIMES is in P, IIT Kanpur. Preprint of August, *8*, 2.
3. Aharoni, R., Erdös, P., & Linial, N., (1985). Dual integer linear programs and the relationship between their optima. In: *Proceedings of the Seventeenth Annual ACM Symposium on Theory of Computing* (pp. 476–483). ACM.
4. Aistleitner, C., (2011). Covering numbers, dyadic chaining and discrepancy. *Journal of Complexity, 27*(6), 531–540.
5. Alkalai, L., & Geer, D., (1996). *Space-Qualified 3D Packaging Approach for Deep Space Missions: New Millennium Program, Deep Space 1 Micro-Electronics Systems Technologies.* Viewgraph presentation. Pasadena, CA: Jet Propulsion Laboratory, 1.
6. Alon, N., & Spencer, J., (2000). *The Probabilistic Method.* With an appendix on the life and work of Paul Erdos. Wiley-Intersci. Ser. Discrete Math. Optim., Wiley-Interscience, New York.
7. Arora, N. S., Blumofe, R. D., & Plaxton, C. G., (2001). Thread scheduling for multiprogrammed multiprocessors. *Theory of Computing Systems, 34*(2), 115–144.
8. Arora, S., (1998). Polynomial time approximation schemes for Euclidean traveling salesman and other geometric problems. *Journal of the ACM (JACM), 45*(5), 753–782.
9. Ausiello, G., Crescenzi, P., Gambosi, G., Kann, V., Marchetti-Spaccamela, A., & Protasi, M., (1999). *Complexity and Approximation·Springer.* Berlin, Heidelberg, New York.
10. Ausiello, G., Crescenzi, P., Gambosi, G., Kann, V., Marchetti-Spaccamela, A., & Protasi, M., (2012). *Complexity and Approximation: Combinatorial Optimization Problems and Their Approximability Properties*. Springer Science & Business Media.
11. Ausiello, G., Marchetti-Spaccamela, A., Crescenzi, P., Gambosi, G., Protasi, M., & Kann, V., (1999). Heuristic methods. In: *Complexity and Approximation* (pp. 321–351). Springer, Berlin, Heidelberg.

12. Becchetti, L., Leonardi, S., Marchetti-Spaccamela, A., Schäfer, G., & Vredeveld, T., (2006). Average-case and smoothed competitive analysis of the multilevel feedback algorithm. *Mathematics of Operations Research, 31*(1), 85–108.
13. Beier, R., & Vöcking, B., (2003). Random knapsack in expected polynomial time. In: *Proceedings of the Thirty-Fifth Annual ACM Symposium on Theory of Computing* (pp. 232–241). ACM.
14. Beier, R., & Vöcking, B., (2004). Probabilistic analysis of knapsack core algorithms. In: *Proceedings of the Fifteenth Annual ACM-SIAM symposium on Discrete Algorithms* (pp. 468–477). Society for Industrial and Applied Mathematics.
15. Beier, R., & Vöcking, B., (2006). Typical properties of winners and losers [0.2 ex] in discrete optimization. *SIAM Journal on Computing, 35*(4), 855–881.
16. Beier, R., Röglin, H., & Vöcking, B., (2007). The smoothed number of pareto optimal solutions in bicriteria integer optimization. In: *International Conference on Integer Programming and Combinatorial Optimization* (pp. 53–67). Springer, Berlin, Heidelberg.
17. Belanger, J., & Wang, J., (1993). Isomorphisms of NP complete problems on random instances. In: *Structure in Complexity Theory Conference, 1993, Proceedings of the Eighth Annual* (pp. 65–74). IEEE.
18. Bennett, E. M., Cramer, W., Begossi, A., Cundill, G., Díaz, S., Egoh, B. N., & Lebel, L., (2015). Linking biodiversity, ecosystem services, and human well-being: Three challenges for designing research for sustainability. *Current Opinion in Environmental Sustainability, 14*, 76–85.
19. Berry, M. V., & Howls, C. J., (2012). *Integrals with Coalescing Saddles. 36*, 775–793.
20. Book, R. V., & Siekmann, J. H., (1986). On unification: Equational theories are not bounded. *Journal of Symbolic Computation, 2*(4), 317–324.
21. Boyd, S. C., & Pulleyblank, W. R., (1990). Optimizing over the subtour polytope of the travelling salesman problem. *Mathematical Programming, 49*(1–3), 163–187.
22. Carlson, J. A., Jaffe, A., & Wiles, A., (2006). *The Millennium Prize Problems*. American Mathematical Soc.

23. Černý, V., (1985). Thermodynamical approach to the traveling salesman problem: An efficient simulation algorithm. *Journal of Optimization Theory and Applications, 45*(1), 41–51.
24. Chvatal, V., (1979). A greedy heuristic for the set-covering problem. *Mathematics of Operations Research, 4*(3), 233–235.
25. Chvatal, V., (1983). *Linear Programming*. Macmillan.
26. Cook, W., & Rohe, A., (1999). Computing minimum-weight perfect matchings. *INFORMS Journal on Computing, 11*(2), 138–148.
27. Cormen, T. H., Leiserson, C. E., Rivest, R. L., & Stein, C., (2001). *Introduction to Algorithms*, Sect. 22.5.
28. Cueto, E., Sukumar, N., Calvo, B., Martínez, M. A., Cegonino, J., & Doblaré, M., (2003). Overview and recent advances in natural neighbor Galerkin methods. *Archives of Computational Methods in Engineering, 10*(4), 307–384.
29. Dantzig, G., (2016). *Linear Programming and Extensions*. Princeton university press.
30. Dotú, I., Del Val, A., & Cebrián, M., (2003). Redundant modeling for the quasigroup completion problem. In: *International Conference on Principles and Practice of Constraint Programming* (pp. 288–302). Springer, Berlin, Heidelberg.
31. Durand, A., Hermann, M., & Kolaitis, P. G., (2005). Subtractive reductions and complete problems for counting complexity classes. *Theoretical Computer Science, 340*(3), 496–513.
32. Edmonds, J., (1965). Maximum matching and a polyhedron with 0, 1-vertices. *Journal of Research of the National Bureau of Standards B, 69*(125–130), 55–56.
33. Feige, U., (2002). Relations between average case complexity and approximation complexity. In: *Proceedings of the Thirty-Fourth Annual ACM Symposium on Theory of Computing* (pp. 534–543). ACM.
34. Feller, W., (1971). *An Introduction to Probability Theory and its Applications* (Vol. 2). Wiley, New York.
35. Festa, P., & Resende, M. G., (2002). GRASP: An annotated bibliography. In: *Essays and Surveys in Metaheuristics* (pp. 325–367). Springer, Boston, MA.
36. Galil, Z., (1974). On some direct encodings of nondeterministic Turing machines operating in polynomial time into P-complete problems. *ACM SIGACT News, 6*(1), 19–24.

37. Garey, M. R., & Johnson, D. S., (1976). Approximation algorithms for combinatorial problems: An annotated bibliography. *Algorithms and Complexity: New Directions and Recent Results,* 41–52.

38. Garey, M. R., Graham, R. L., & Ullman, J. D., (1972). Worst-case analysis of memory allocation algorithms. In: *Proceedings of the fourth annual ACM Symposium on Theory of Computing* (pp. 143–150). ACM.

39. Geman, S., & Geman, D., (1987). Stochastic relaxation, Gibbs distributions, and the Bayesian restoration of images. In: *Readings in Computer Vision* (pp. 564–584).

40. Goemans, M. X., & Williamson, D. P., (1995). Improved approximation algorithms for maximum cut and satisfiability problems using semidefinite programming. *Journal of the ACM (JACM), 42*(6), 1115–1145.

41. Gomes, C. P., & Shmoys, D. B., (2002). The promise of LP to boost CSP techniques for combinatorial problems. In: *Proc., Fourth International Workshop on Integration of AI and OR techniques in Constraint Programming for Combinatorial Optimization Problems (CP-AI-OR'02), Le Croisic, France* (pp. 25–27).

42. Gomes, C. P., Selman, B., & Kautz, H., (1998). Boosting combinatorial search through randomization. *AAAI/IAAI, 98*, 431–437.

43. Gomes, C. P., Selman, B., Crato, N., & Kautz, H., (2000). Heavy-tailed phenomena in satisfiability and constraint satisfaction problems. *Journal of Automated Reasoning, 24*(1, 2), 67–100.

44. Gomes, C., & Shmoys, D., (2002). Completing quasigroups or Latin squares: A structured graph coloring problem. In: *Proceedings of the Computational Symposium on Graph Coloring and Generalizations* (pp. 22–39).

45. Gurevich, Y., (1990). Matrix decomposition problem is complete for the average case. In: *Foundations of Computer Science, 1990. Proceedings, 31st Annual Symposium* (pp. 802–811). IEEE.

46. Gurevich, Y., (1991). Average case complexity. In: *International Colloquium on Automata, Languages, and Programming* (pp. 615–628). Springer, Berlin, Heidelberg.

47. Halperin, E., (2002). Improved approximation algorithms for the vertex cover problem in graphs and hypergraphs. *SIAM Journal on Computing, 31*(5), 1608–1623.

48. Hermann, M., & Kolaitis, P. G., (1994). The complexity of counting problems in equational matching. In: *International Conference on Automated Deduction* (pp. 560–574). Springer, Berlin, Heidelberg.
49. Hermann, M., & Pichler, R., (2008). Complexity of counting the optimal solutions. In: *International Computing and Combinatorics Conference* (pp. 149–159). Springer, Berlin, Heidelberg.
50. Ho, A. C., (1982). Worst case analysis of a class of set covering heuristics. *Mathematical Programming, 23*(1), 170–180.
51. Hochbaum, D. S., & Shmoys, D. B., (1987). Using dual approximation algorithms for scheduling problems theoretical and practical results. *Journal of the ACM (JACM), 34*(1), 144–162.
52. Hochbaum, D. S., (1996). *Approximation Algorithms for NP-Hard Problems*. PWS Publishing Co.
53. Indrani, A. V., (2003). *Some issues concerning Computer Algebra in AToM3*. Technical Report, School of Computer Science, McGill University, Montreal, QC.
54. Jain, K., & Vazirani, V. V., (2001). Approximation algorithms for metric facility location and k-median problems using the primal-dual schema and lagrangian relaxation. *Journal of the ACM (JACM), 48*(2), 274–296.
55. Jiménez, P., Thomas, F., & Torras, C., (2001). 3D collision detection: A survey. *Computers & Graphics, 25*(2), 269–285.
56. Johnson, D. S., (1973). Approximation algorithms for combinatorial problems. In: *Proceedings of the Fifth Annual ACM Symposium on Theory of Computing* (pp. 38–49). ACM.
57. Karp, R. M., (1975). The fast approximate solution of hard combinatorial problems. In: *Proc. 6th South-Eastern Conf. Combinatorics, Graph Theory and Computing (Florida Atlantic U. 1975)* (pp. 15–31).
58. Khot, S. A., & Vishnoi, N. K., (2015). The unique games conjecture, integrality gap for cut problems and embeddability of negative-type metrics into ℓ 1. *Journal of the ACM (JACM), 62*(1), 8.
59. Khot, S., & Regev, O., (2003). Vertex cover might be hard to approximate to within 2-/spl epsiv. In: *Computational Complexity, 2003; Proceedings 18th IEEE Annual Conference* (pp. 379–386). IEEE.

60. Khot, S., (2002). On the power of unique 2-prover 1-round games. In: *Proceedings of the Thirty-Fourth Annual ACM symposium on Theory of Computing* (pp. 767–775). ACM.
61. Khot, S., Kindler, G., Mossel, E., & O'Donnell, R., (2007). Optimal inapproximability results for MAX-CUT and other 2-variable CSPs?. *SIAM Journal on Computing, 37*(1), 319–357.
62. Klein, P. N., & Young, N. E., (2010). Approximation algorithms for NP-hard optimization problems. In: *Algorithms and Theory of Computation Handbook* (pp. 34–34). Chapman & Hall/CRC.
63. Kozen, D. C., (1992). Counting problems and# P. In: *The Design and Analysis of Algorithms* (pp. 138–143). Springer, New York, NY.
64. Kumar, S. R., Russell, A., & Sundaram, R., (1999). Approximating Latin square extensions. *Algorithmica, 24*(2), 128–138.
65. LaForge, L. A., & Turner, J. W., (2006). Multi-processors by the numbers: Mathematical foundations of spaceflight grid computing. In: *Aerospace Conference, 2006 IEEE* (p. 19). IEEE.
66. LaForge, L. E., Moreland, J. R., & Fadali, M. S., (2006). Spaceflight multi-processors with fault tolerance and connectivity tuned from sparse to dense. In: *Aerospace Conference, 2006 IEEE* (p. 23). IEEE.
67. Laywine, C. F., & Mullen, G. L., (1998). *Discrete Mathematics Using Latin Squares* (Vol. 49). John Wiley & Sons.
68. Leahu, L., & Gomes, C. P., (2004). Quality of LP-based approximations for highly combinatorial problems. In: *International Conference on Principles and Practice of Constraint Programming* (pp. 377–392). Springer, Berlin, Heidelberg.
69. Lenstra, J. K., Shmoys, D. B., & Tardos, E., (1990). Approximation algorithms for scheduling unrelated parallel machines. *Mathematical Programming, 46*(1–3), 259–271.
70. Liu, C. L., (1976). Deterministic job scheduling in computing systems. In: *Performance* (pp. 241–254).
71. Mossel, E., O'Donnell, R., & Oleszkiewicz, K., (2005). Noise stability of functions with low influences: Invariance and optimality. In: *Foundations of Computer Science, 2005, FOCS 2005; 46th Annual IEEE Symposium* (pp. 21–30). IEEE.
72. Motwani, R., & Raghavan, P., (1995). *Randomized Algorithms*. Cambridge International Series on Parallel Computation.

73. Nowakowski, A., & Skarbek, W., (2006). Fast computation of thresholding hysteresis for edge detection. In: *Photonics Applications in Astronomy, Communications, Industry, and High-Energy Physics Experiments IV* (Vol. 6159, p. 615948). International Society for Optics and Photonics.
74. Paz, A., & Moran, S., (1977). Non-deterministic polynomial optimization problems and their approximation. In: *International Colloquium on Automata, Languages, and Programming* (pp. 370–379). Springer, Berlin, Heidelberg.
75. Podsakoff, N. P., Whiting, S. W., Podsakoff, P. M., & Blume, B. D., (2009). Individual-and organizational-level consequences of organizational citizenship behaviors: A meta-analysis. *Journal of applied Psychology, 94*(1), 122.
76. Pulleyblank, W. R., (1989). Chapter V; Polyhedral combinatorics. *Handbooks in Operations Research and Management Science, 1*, 371–446.
77. Qi, L., (1988). Directed submodularity, ditroids and directed submodular flows. *Mathematical Programming, 42*(1–3), 579–599.
78. Reinelt, G., (1994). *The Traveling Salesman: Computational Solutions for TSP Applications*. Springer-Verlag.
79. Shaw, P., Stergiou, K., & Walsh, T., (1998). Arc consistency and quasigroup completion. In: *Proceedings of the ECAI-98 Workshop on Non-Binary Constraints* (Vol. 2).
80. Shmoys, D. B., (1995). Computing near-optimal solutions to combinatorial optimization problems. *Combinatorial Optimization, 20*, 355–397.
81. Stock, J. H., & Watson, M. W., (2001). Vector autoregressions. *Journal of Economic Perspectives, 15*(4), 101–115.
82. Vangheluwe, H., Sridharan, B., & Indrani, A. V., (2003). *An Algorithm to Implement a Canonical Representation of Algebraic Expressions and Equations in AToM3*. Technical Report, School of Computer Science, McGill University, Montreal, QC.
83. Vazirani, V. V., (2013). *Approximation Algorithms*. Springer Science & Business Media.
84. Vershynin, R., (2009). Beyond Hirsch conjecture: Walks on random polytopes and smoothed complexity of the simplex method. *SIAM Journal on Computing, 39*(2), 646–678.

85. Wang, Y., (2008). Topology control for wireless sensor networks. In: *Wireless Sensor Networks and Applications* (pp. 113–147). Springer, Boston, MA.
86. Wilkinson, M. H., (2003). Gaussian-weighted moving-window robust automatic threshold selection. In: *International Conference on Computer Analysis of Images and Patterns* (pp. 369–376). Springer, Berlin, Heidelberg.
87. Williams, R., Gomes, C. P., & Selman, B., (2003). Backdoors to typical case complexity. In: *IJCAI* (Vol. 3, pp. 1173–1178).
88. Xu, W. S., (2005). *The Design and Implementation of the μModelica Compiler*. Doctoral dissertation, MSc. Thesis. School of Computer Science, McGill University, Montreal, QC.
89. Zhu, X., & Wilhelm, W. E., (2006). Scheduling and lot sizing with sequence-dependent setup: A literature review. *IIE Transactions, 38*(11), 987–1007.

CHAPTER 8

GOVERNANCE OF ALGORITHMS

CONTENTS

8.1. INTRODUCTION

The purpose of this chapter is to provide a contribution to the development of a greater grasp of governance choice in the context of algorithm selection. Algorithms on the Internet have a role in the creation of our realities and the conduct of our everyday lives. It is their job to first choose the material, then mechanically assign significance to it, saving individuals from drowning in a sea of knowledge. Nonetheless, the benefits of algorithms come with a number of hazards as well as governance difficulties to consider.

According to assessments of actual case studies and a literature study, we shall outline a risk-based approach to corporate governance. This technique analyzes and then categorizes the applications of algorithmic choice, as well as the dangers associated with them. Following that, it investigates the wide range of institutional governance alternatives available and briefly analyzes the many governance measures that have been implemented and recommended for algorithmic selection, as well as the constraints of governance options. According to the findings of the study, there are no one-size-fits-all methods for regulating algorithms (Figure 8.1).

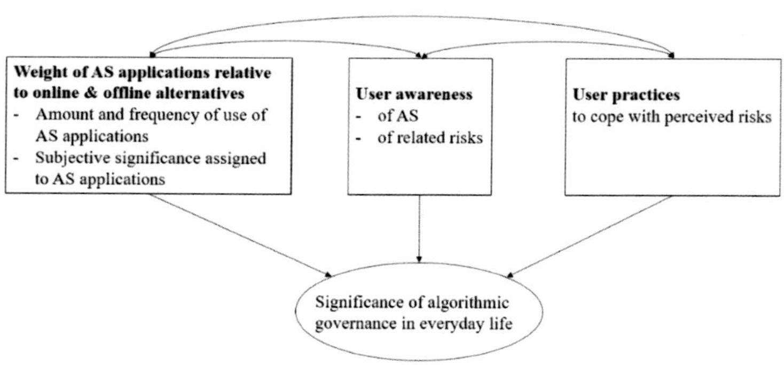

Figure 8.1. The theoretical model of variables measuring the significance of algorithmic governance in everyday life.

Note: AS: Algorithmic selection.

A growing number of algorithms have been implemented into the Internet-based apps that we use in our everyday lives. These software intermediaries operate in the background and have an impact on a wide range of operations. The ingestion of video and music entertainment through recommender systems, the selection of online news through various news and aggregators search engines, the display of status messages on various online social networks, the selection of products and services in online shops, and algorithmic trading in stock exchange markets worldwide are some of the most visible examples of this pervasive trend. While their purpose and operation modes greatly differ in detail, Latzer et al. (2015) identified nine groups of different Internet services that are dependent on algorithmic selection applications because while their purpose and operation modes greatly differ in feature, all of these applications are categorized by a mutual basic functionality; they all automatically select information elements (Senecal and Nantel, 2004; Hinz and Eckert, 2010).

The widespread proliferation of algorithms in an increasing number of fields is one of the key explanations for the rising discussion on the "power of algorithms." The influence of recommendation systems on customer choice in electronic commerce, the effect of Google rankings (Epstein and Robertson, 2013; Döpfner, 2014), and the impact of Facebook's News Feed on the news industry are all examples of this power (Table 8.1) (Bucher, 2012; Somaiya, 2014).

Table 8.1. Illustration of Different Algorithm Types and Their Examples

Types	Examples
Search	· General search engines · Metasearch engines · Special search engines · Question and answer services · Semantic search engines
Allocation	· Computational advertising (e.g., Google AdSense, Yahoo! Bing Network) · Algorithmic trading (e.g., Quantopian)
Scoring	· Reputation systems: music, film, etc. (e.g., eBay's reputation system) · News scoring (e.g., Reddit, Digg) · Credit scoring (e.g., Kreditech) · Social scoring (e.g., Klout)
Content production	· Algorithmic journalism (e.g., Quill; Quakebot)
Recommendation	· Recommender systems (e.g., Spotify; Netflix)
Filtering	· Spam filter (e.g., Norton) · Child protection filter (e.g., Net Nanny)
Prognosis/forecast	· Predictive policing (e.g., PredPol) · Predicting developments: success, diffusion, etc. (e.g., Google Flu Trends, scoreAhit)

Observation/ surveillance	· Surveillance · Employee monitoring · General monitoring software
Aggregation	· News aggregators

The dominance of Facebook's and Google's algorithms stands out in a broader discussion over the economic and social consequences of software in common and algorithms in specific and serves as a notable example. Software, as per Manovich (2013), "takes command" by substituting a diverse array of mechanical, physical, and electrical technologies that are responsible for the creation, allocation, supply, and interaction with cultural objects, among other things (Musiani, 2013; Gillespie, 2014; Pasquale, 2015). In the same way that laws and regulations have controlling powers, algorithms, and codes have as well (Lessig, 1999; Mager, 2012). "The power of technology" and "Increasing automation" have been extensively discussed by a number of journalists and researchers whose primary focus is the role of algorithms and code as agents, ideologies, institutions, gatekeepers, as well as modes of intermediation and modes of mediation (Machill and Beiler, 2007; Steiner, 2012). From the standpoint of intermediation, their role as receptionists and their effect on the development of public opinion, the establishment of public circles, and the building of realities are particularly highlighted. In information societies, algorithmic selection automates a profit-driven reality-construction and reality-mining process (Jürgens et al., 2011; Katzenbach, 2011; Napoli, 2013; Wallace and Dörr, 2015).

8.2. ANALYTICAL FRAMEWORK

Dangers give grounds for algorithm control; these risks originate from the distribution of algorithmic selection. Governance should reinforce benefits while reducing dangers from a public-interest standpoint (Van Dalen, 2012). Advantages and risks are inextricably linked since hazards jeopardize advantage exploitation. As a result, a "risk-based strategy" (Black, 2010) analyzes and analyzes the risks, as well as the possibilities and constraints for reducing them.

Latzer et al. (2007) have identified nine classes of risk which be associated with algorithmic selection (Figure 8.2):

- Manipulation (Rietjens, 2006; Bar-Ilan, 2007; Schormann, 2012);

- Restrictions on the freedom of expression and communication, for instance, censorship through intelligent filtering (Zittrain and Palfrey, 2008);
- Weakening variety, the formation of echo chambers (Bozdag, 2013) and filter bubbles (Pariser, 2011), biases and distortions of reality;
- Surveillance and threats to data protection and privacy;
- Social discrimination;
- Violation of intellectual property rights;
- Abuse of market power;
- Effects on cognitive capabilities and the human brain;
- Growing heteronomy and loss of human sovereignty and controllability of technology.

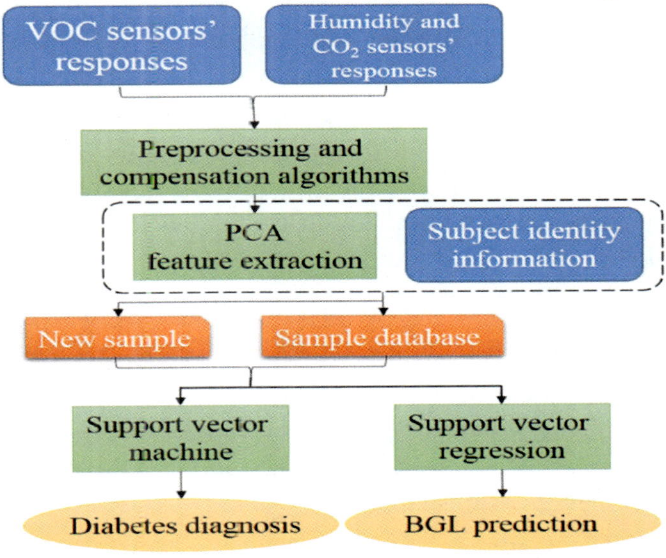

Figure 8.2. The framework of the data analysis algorithms. The rounded rectangles in blue are the features. The rectangles in green are the algorithms.

Source: https://www.researchgate.net/figure/Framework-of-the-data-analysis-algorithms-The-rounded-rectangles-in-blue-are-the_fig1_263288691.

There are a variety of governance methods available for algorithmic selection that may help to reduce risks while also enhancing profitability. As a result of having various types of resources at their disposal, different actors take different techniques and, as a result, have varying levels of skill. A "governance viewpoint," as it is often understood by scientists, is a valuable lens through which to analyze, assess, and improve regulatory policies (Grasser and Schulz, 2015). If we look at governance from an institutional standpoint, there is a continuum that runs from market contrivances on one end of the spectrum to control and command regulation via state authorities on the other end of the spectrum. The middle ground is occupied by a variety of alternate governance modes, which fall into the categories of (2) self-help through single companies; (3) collective self-regulation with the assistance of industry branches; and (4) co-regulation, which is a regulatory collaboration between industry state and authorities, among others.

For some years now, researchers have been paying close attention to alternate methods to governance (Gunningham and Rees, 1997; Sinclair, 1997), and their findings have been widely published. In particular, their applicability, application, as well as performance in the communications industries are all being investigated. The majority of the time, in a market economy, market solutions are favored over government involvement. However, only when issues cannot be resolved by private activity (subsidiarity) is it necessary for the state to intervene. Nevertheless, it must be justified by the alleged restrictions or failures of market solutions, as well as by the self-regulation of the industry in question. In order to do so, we must make a comparison between the advantages and disadvantages of alternative governance models.

There are two significant pillars that form the basis of this assessment:

- It is informed by proof of risk-specific measures of governance, comprising previously established and hitherto only proposed interventions. Overall, this exhibits an extensive range of governance options; and
- The evaluation, moreover, rests on a structure for choice of governance (Saurwein, 2011).

For algorithmic selection, there is a range of governance mechanisms that can assist decrease risks while simultaneously increasing profitability. Different actors use different tactics as a consequence of having different sorts of materials at their disposal, and as a result, they have differing degrees of expertise. A "governance viewpoint," as it is commonly referred to by scientists, is a useful lens for analyzing, evaluating, and improving regulatory systems (Grasser and Schulz, 2015). When it comes to institutional governance, there is a continuum that goes from (1) market manipulations on one end of the spectrum to (5) control and command regulation by state authorities on the other. Various different governance systems occupy the middle ground, including (2) self-help through single firms; (3) collective self-regulation with the support of industry sections; and (4) co-regulation, which is a regulatory partnership between state and industry authorities, among others (Bartle and Vass, 2005).

Researchers have been studying alternative governance techniques for some years (Gunningham and Rees, 1997; Sinclair, 1997), and their findings have been extensively disseminated. Their adaptability, use, and performance in the communications sectors, in particular, are all being researched. In a market economy, market solutions are usually preferred above government intervention. However, the state must interfere only when difficulties cannot be remedied through private action. However, it must be explained by the apparent failures or limitations of market solutions, as well as by the industry's self-regulation. To do so, we must compare the benefits and drawbacks of various governance structures.

8.3. GOVERNANCE OPTIONS BY RISKS

We investigate the governance of algorithms by doing a positive analysis of the mechanisms that have been created or those that have been proposed for managing the hazards associated with algorithmic selection. As a result of the investigation, we have a summary of patterns in the governance of algorithms, which reveals distinct disparities in the selection and grouping of different governance techniques in response to certain hazards.

Some hazards have previously been handled by various ways to governance (data protection), whilst others have yet to be addressed through any means (heteronomy). Certain risks are often left entirely to market solutions (bias). However, for a few others, governance is institutionalized equally by both state and private regulatory mechanisms, resulting in a hybrid governance structure. Though several arrangements and proposals for measures in the form of self-organization through corporations exist, there are only a small number of co-regulatory arrangements in place. Overall, there does not appear to be any overarching institutional framework for managing the hazards associated with algorithmic selection. In spite of this, there is a wide range of currently implemented governance measures, in addition to proposals by policymakers and scholars for more governance measures.

8.3.1. Potential of Market Solutions and Governance by Design

For all of the dangers connected with algorithmic selection, specific governance mechanisms are not required. We may also lower risks by changing the market behavior of content providers, consumers, and algorithmic service providers voluntarily. Consumers, for example, might avoid utilizing problematic services by transferring to another supplier or relying on knowledge to protect themselves against hazards. Consumers can benefit from technological self-help solutions, for example. Bias, censorship, and privacy infringement are reduced as a result of such solutions. Clients can use a variety of anonymization techniques, like virtual private networks (VPNs), Tor, or Open DNS, to avoid censorship and protect their privacy. Cookie management, encryption, and do-not-track technologies are examples of privacy-enhancing technologies (PETs) that may be used to secure data (browser). As a result, using chances for de-personalization of services can help to eliminate prejudice. In general, these examples show several choices for user self-protection, although several of these "demand-side solutions" are reliant on and reduced by the availability of sufficient supply (Cavoukia, 2012).

Providers of such services, which are based on algorithmic selection, might mitigate risks by employing commercial tactics. This can be accomplished through introducing product innovations, such as updates to existing services or the introduction of new services. There are various instances of such services that have been developed to avoid copyright and privacy infringement and prejudice. Few news aggregators' business models include content suppliers who are compensated (for example, nachrichten.de). It is possible to minimize privacy problems by using algorithmic services that do not gather user data (Resnick et al., 2013; Krishnan et al., 2014). If these product developments are successful, they may contribute to market variety as well as a decrease in market concentration (Schaar, 2010).

Other examples focus on the technology layout of services for mitigating risks, such as prejudice, privacy violations, and manipulation. "Privacy by design" and "Privacy by default" are two technological methods to enhance privacy. By including serendipity components, services like Reflect, ConsiderIt, and OpinionSpace aim to eliminate prejudice and filter bubbles (Munson and Resnick, 2010; Schedl et al., 2012). Machine learning can significantly minimize bias in recommender systems. Strong self-protection is, therefore, for the suppliers of algorithmic services' own good in order to avoid manipulation. They frequently employ technical safeguards to counteract third-party exploitation. We can see a digital arms race in areas such as filtering, recommendation, and search, where content producers are putting out the effort to prevent problems by applying content-optimization tactics (Jansen, 2007; Wittel and Wu, 2004). Copyright infractions are also avoided through technical self-help (robots.txt files) by content producers.

8.3.2. Options for the Industry: Self-Regulation and Self-Organization

Individual enterprises can minimize costs through "self-organization" in addition to technical self-protection and product improvements. Company norms and principles that represent internal quality evaluation in respect to specified hazards, the public interest, and ombudsman programs for dealing with complaints are typical instances of self-organization (Langheinrich, 2001; Cavoukia, 2009). The dedication to self-organization is usually part of a company's overall CSR (corporate social responsibility) strategy. From an economic standpoint, the goal of self-organization is to improve or avoid losing reputation. Service providers whose services are based on algorithmic selection may give to certain "values," like the "minimal principle" of data acquisition. There are also different recommendations for ethical boards at the corporate level to deal with involvement in user experiences or software development concerns (Lin and Selinger, 2014). Companies can create principles as well as observe internal quality control for various hazards such as discrimination, prejudice, and manipulation. Google, for instance, announced the formation of an ethics board. In the context of big data, in-house algorithms have been proposed as a way to oversee big-data operations and as the first point of contact for individuals who feel misled by an organization's big-data activities (Mayer-Schönberger and Cukier, 2013).

Unlike individual company self-organization, self-regulation refers to a group of companies/branches working together to achieve public-interest goals through self-restriction. Technical and organizational industry standards, codes of conduct, arbitration, and ombudsmen boards, quality seals and certification agencies, and ethics committees/commissions are only some examples of industry self-regulation instruments. There are sectoral self-regulation projects in the marketing business (Europe, USA), online social networks, the search engine market, and algorithmic trading in an extended field of algorithmic selection. These initiatives address issues like copyright and privacy infringement, controllability, and algorithmic transaction manipulation. The stock exchange has implemented warning and monitoring systems to identify manipulation and circumstances when automated trading runs out of hand. Similarly, in the field of online behavioral advertising, there are a number of attempts in the advertising business to improve data privacy (OBA). The Digital Advertising Alliances in the United States and Europe are in charge of this. Various tools, including codes of conduct, general online opt-out boundaries for customers, and certification systems, are part of the projects. In addition, the advertising industry is active in the technological standards for do-not-track, alongside Web browser providers (DNT). Furthermore, industry efforts such as digital rights management systems (DRM) and the creative commons licensing system exist to preserve copyrights on a technological and organizational level. In this instance, "self-regulation" through shared standards tailored to the interests of the business would be appropriate. Furthermore, ombudsmen, ethics commissions, and certification systems appear to be appropriate tools for dealing with the specific hazards of algorithmic selections. Nonetheless, the sector has yet to implement these choices, and it appears that there is a significant amount of untapped potential for self-regulatory governance solutions.

8.3.3. Examples and Possibilities of State Intervention

The algorithmic selection also poses issues to the state and to political institutions, as previously stated. The limits of market systems, as well as the effectiveness of self-regulation in reducing risks, can serve as explanations and arguments for government involvement. Command-and-control regulation, the provision of public services, enticements in the form of taxes/fees and subsidies/funding, co-regulation, information measures, and soft law are all examples of typical state intervention instruments (Lewandowski, 2014). These instruments are used to increase people's knowledge and awareness of risks in order to encourage appropriate behavior. It is possible to find multiple examples of governmental action in the sphere of algorithmic selection; furthermore, the restrictions are linked to specific hazards rather than a specific technology or a specific industry (Schulz et al., 2005). Individuals in Europe, for example, are protected against automated choices on some personal elements such as work performance, reliability, creditworthiness, and conduct under the European Union's (EU) privacy legislation (Argenton and Prüfer, 2012). Another area of ongoing regulatory dispute is the use of search engines on the internet. Google is being investigated by European and American competition authorities due to worries about fair competition.

Contestants allege that a search on Google offers an unwarranted advantage to the company's other services, prompting the authorities to launch an inquiry. The majority of proposals for governing activities in the search-engine market call for increased controllability and transparency on the part of public authorities, while only a minority of proposals seek to reduce market entry barriers or establish the principle of neutral search. In order to promote market contestability, aid market entrance, and support healthy competition, it is advocated that the public fund an "index of the web" or user data sets as shared resources (Lao, 2013).

In addition to control-and-command regulation, state actors can employ different modes of interference, like taxes, soft law, subsidies, co-regulation, and information, in addition to control-and-command regulation. The implementation of a machine tax to offset financial losses caused by automation, as well as the introduction of a data tax/fee to reduce the economic incentives for data collecting, have been advocated by a few. In most cases, the state intervenes through the use of monetary incentives. For example, there are a number of initiatives aimed at maximizing the potential of automation in the manufacturing business by encouraging reorganization in the sector. However, financing can also be utilized to assist in the reduction of risk. For instance, the EU funds the growth of PETs through its Research and Development initiatives. In the realm of data protection, co-regulation, and soft law have also become established practices. Renowned instruments include certification schemes for data protection and quality assurance seals, as well as the Safe Harbor Principles and the Fair Information Practice Principles in the United States, which control data transfers for commercial purposes between the EU and the United States (Collin and Colin, 2013).

8.4. LIMITATIONS OF ALGORITHMIC GOVERNANCE OPTIONS

The discovery of algorithms' powerful influence ("government by algorithms") has sparked a debate about how to properly control these capabilities ("governance of algorithms"). Google's dominating and influential position, in particular, is regularly challenged (Zuboff, 2014). The administration of internet search is gaining public and regulatory attention (Lewandowski, 2014; König and Rasch, 2014). Disagreements over some search and news aggregation tactics and outcomes have been caused in regulatory legislation addressing privacy and copyright infringement. The German auxiliary copyright is one example (Figure 8.3) (Moffat, 2009).

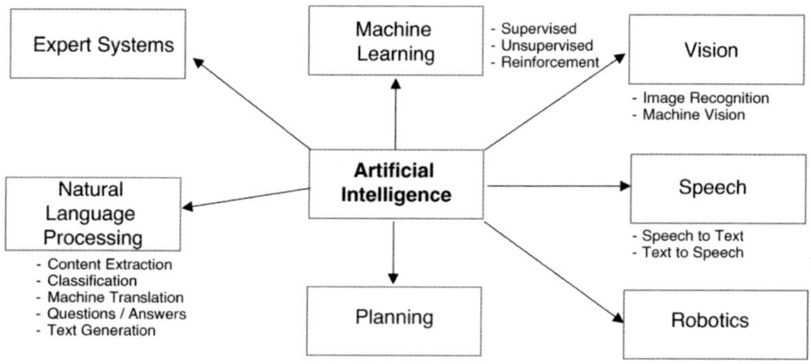

Figure 8.3. Algorithmic decision systems.

Source: https://www.cambridge.org/core/books/abs/cambridge-handbook-of-the-law-of-algorithms/algorithmic-decision-systems/B5731E525B19EB-D98B132CC20A0DD7F6.

Nonetheless, algorithmic selection's uses and related concerns extend far beyond internet search and Google. As a result, it is critical to broadening the area of study in order to appreciate the broad range of applications, their function, related features, and repercussions for societies and markets, as well as their numerous problematic implications and governance prospects (Latzer, 2007, 2014). This chapter briefly examines algorithmic governance by offering the rationales for algorithmic governance as well as an outline and classification of the hazards associated with algorithmic selection. There are various institutional governance choices accessible, each with its own set of restrictions. We may make a conclusion for our choice of governance based on our tastes and requirements by considering the limits of governance alternatives (Latzer et al., 2015).

In addition to the range of governance measures, the choice of governance must take into account the institutional governance alternatives' boundaries. Contextual variables for governance help to describe the possibility of implementing specific governance systems as well as their suitability in respect to certain hazards (Saurwein, 2011). Options are implicated by the consideration of enabling contextual elements following the boundaries/limitations of institutional governance.

8.4.1. Limitations of Market Solutions and Self-Help Strategies

Consumer self-help measures (switch, opt-out, technological self-protection) can assist in mitigating some of the hazards of algorithmic selection;

nonetheless, there are a number of roadblocks to successful self-help; additionally, we must not overstate the capability of user self-protection. Consumers have the option of discontinuing troublesome services or switching to different goods. Algorithmic applications, on the other hand, frequently operate without express agreement. For example, there is no way to opt out of a government monitoring program. Switching service providers need the availability of replacement services; yet, many markets are highly consolidated, and hence switching chances are restricted (Lao, 2013). Due to information asymmetries, the hazards of algorithmic selection are frequently overlooked by customers, resulting in poor-risk awareness. A typical Internet user, for example, is unlikely to notice censorship, manipulation, or bias. As a result, if hazards are not obvious, there is no incentive to seek self-protection techniques (Langheinrich, 2001). Free services, on the other hand, reduce transparency and provide consumers with fewer incentives to switch to lower-risk alternatives. If there are technological instruments for self-defense accessible, they almost often require abilities that the majority of users lack. For example, in the realm of data protection, anonymization necessitates technological expertise and may be challenged by succeeding re-identification. Ultimately, using self-protection or switching tactics needs the availability of alternative services and protective technology. As a result, in terms of access to tools and services, customer options are decided by the supply side of the market (Ohm, 2010).

Another way to mitigate the risks of algorithmic selection is to implement supply-side measures (for example, product improvements), but suppliers, too, suffer constraints when it comes to risk-reduction business strategies. First and foremost, certain market categories have substantial entry barriers, making circumstances tough for newcomers and product improvements. Furthermore, risk minimization might lead to a drop in service quality, resulting in competitive drawbacks. For instance, while services devoid of personalization decrease the potential of privacy violations, they may also reduce the value of the service for users. As a result, "alternative goods" are frequently specialized services with a small customer base. Reduced quality and a small number of consumers reinforce each other, lowering the attraction of specialized services even further.

8.4.2. Limitations of Self-Regulation and Self-Organization

For self-organization, the examination of governance measures at the firm level reveals several choices; however, impediments obstruct voluntary approaches. Implementation is frequently based on incentives, which are the cost and benefits to the firm (London Economics, 2010; Hustinx, 2010). For example, there may be no incentive for robust voluntary standards in the domain of data privacy. Data has been referred to as the "new oil" in the 21st century. As a result, it is a necessary source for both service innovation and economic success. As a result, it's improbable that businesses will readily stop collecting data. Several governance solutions aim to improve algorithmic process transparency (Elgesem, 2008). Companies, on the other hand, have little incentive to freely publish algorithms because doing so increases the risk of copying and manipulation. As a result, a "transparency challenge" has arisen (Rieder, 2005; Bracha and Pasquale, 2008; Granka, 2010). Furthermore, a company's willingness to self-organize is influenced by its reputation sensitivity (Latzer et al., 2007). Increased focus on firms in the B2C (business-to-consumer) market, like Amazon, may encourage self-restriction in the public interest. On the other side, reduced public emphasis on B2B enterprises like data brokers reduces reputation sensitivity and, as a result, the reasons for voluntary self-organization (Krishnan et al., 2014).

The examination of current governance methods reveals a few examples of industrial branches cooperating to regulate themselves (for instance, advertising). In actuality, the activities are limited to specific hazards in well-established and narrowly defined industries, but the overall background requirements for self-regulation are complicated. Self-regulation is hindered most notably by the variety and fragmentation of the sectors concerned. Advertising, news, entertainment, social interaction, commerce, health, and traffic are just a few of the industries where algorithmic selection is used (Lessig, 1999; Jürgens et al., 2011; Katzenbach, 2011). A broad self-regulatory initiative is unlikely due to the great number and variability of the branches. Furthermore, due to the diversity of the businesses involved, self-regulatory solutions for the lowest of standards are not likely. As a result, statutory regulation must be used to establish basic standards that apply to all market participants (Jansen, 2007). Aside from heterogeneity, there are a few other variables that make self-regulation difficult. Self-regulation, for instance, is more expected to occur in established sectors with like-minded market participants. However, the markets for services that depend on the algorithmic selection are frequently experimental and new (for example, algorithmic content generation), and the algorithmic solution developers are typically novices looking to disrupt established business models and market structures. Newcomers are often on the lookout for fresh opportunities and, as a result, do not always comply with existing industry strategies (Hinz and Eckert, 2010).

8.4.3. Limitations of State Intervention

Ultimately, the study of governance possibilities points to a broad spectrum of prospects for state action in order to mitigate the dangers associated with algorithmic selection, which is encouraging. However, when it comes to the governance of algorithms, the state is not exempt from these constraints. In general, neither every form of risk is well-suited to state intervention, nor is control in particular (Gunningham and Rees, 1997; Grasser and Schulz, 2015). It is difficult for legislative instructions to address risks like cognitive impacts, prejudice, and heteronomy of algorithmic selection since they are so complex. Several examples testify to a lack of practicability and legitimacy in the case of government involvement. For example, in the event of bias issues, the goal of enhancing "objectivity" can be pursued in order to mitigate the problem (Epstein and Robertson, 2013). Aside from that, because several markets are still in their early stages, there is only a limited amount of knowledge available about the future growth of markets, as well as the dangers that may be associated with them. The uncertainty is exacerbated by the fact that threats like "uncontrollability" are unique and that there is little prior experience with issues of a similar kind to those being faced. In addition, because of the complex interdependencies that exist within the socio-technical system, the impact of possible state regulatory measures is usually difficult to foresee. Because of persistent ambiguities surrounding the growth of a market or the impact of regulatory policy, the control of algorithmic selection has been hampered, and as a result, the role of the state has not been decided yet (Gillespie, 2014).

REFERENCES

1. Argenton, C., & Prüfer, J., (2012). Search engine competition with network externalities. *Journal of Competition Law & Economics, 8*(1), 73–105.
2. Bar-Ilan, J., (2007). Google bombing from a time perspective. *Journal of Computer-Mediated Communication, 12*(3), 910–938.
3. Bartle, I., & Vass, P., (2005). *Self-Regulation and the Regulatory State: A Survey of Policy and Practice* (Vol. 1, pp. 1–22). University of Bath School of Management.
4. Black, J., (2010). Risk-based regulation: Choices, practices and lessons learnt. In: OECD, (ed.), *Risk and Regulatory Policy: Improving the Governance of Risk* (pp. 185–224). OECD Publishing, Paris.
5. Bozdag, E., (2013). Bias in algorithmic filtering and personalization. *Ethics and Information Technology, 15*(3), 209–227.
6. Bracha, O., & Pasquale, F., (2008). Federal search commission? Access, fairness and accountability in the law of search. *Cornell Law Review, 93*(6), 1149–1210.
7. Bucher, T., (2012). Want to be on top? Algorithmic power and the threat of invisibility on Facebook. *New Media & Society, 14*(7), 1164–1180.
8. Cavoukia, A., (2012). *Privacy by Design: Origins, Meaning, and Prospects for Ensuring Privacy and Trust in the Information Era* (Vol. 1, pp. 1–20). Available at: https://www.researchgate.net/publication/304011793_Tortius_Privacy_30_a_quest_for_research (accessed on 4 August 2022).
9. Collin, P., & Colin, N., (2013). *Expert Mission on the Taxation of the Digital Economy* (Vol. 1, pp. 1–20). Available at: www.redressement-productif.gouv.fr/files/rapport-fiscalite-du-numerique_2013.pdf (accessed on 4 August 2022).
10. Döpfner, M., (2014). "Warum wir Google fürchten: Offener brief an Eric Schmidt", Frankfurter Allgemeine Zeitung (Vol. 1, pp. 1–19). Available at: www.faz.net/aktuell/feuilleton/medien/mathias-doepfnerwarum-wir-google-fuerchten-12897463.html (accessed on 4 August 2022).
11. Elgesem, D., (2008). Search engines and the public use of reason. *Ethics and Information Technology, 10*(4), 233–242.
12. Epstein, R., & Robertson, R. E., (2013). "*Democracy at Risk: Manipulating Search Rankings Can Shift Voters' Preferences*

Substantially Without Their Awareness (Vol. 1, pp. 1–20). Available at: http://aibrt.org/downloads/EPSTEIN_and_Robertson_2013-Democracy_at_Risk-APS-summary-5-13.pdf (accessed on 4 August 2022).

13. Gillespie, T., (2014). The relevance of algorithms. In: Gillespie, T., Boczkowski, P., & Foot, K., (eds.), *Media Technologies: Essays on Communication, Materiality, and Society* (pp. 167–194). MIT Press, Cambridge, MA.
14. Granka, L. A., (2010). The politics of search: A decade retrospective. *The Information Society, 26*(5), 364–374.
15. Grasser, U., & Schulz, W., (2015). *Governance of Online Intermediaries Observations from a Series of National Case Studies* (Vol. 1, pp. 1–23). Available at: https://dash.harvard.edu/bitstream/1/16140636/1/Berkman_2015-5_final.pdf (accessed on 4 August 2022).
16. Gunningham, N., & Rees, J., (1997). Industry self-regulation: An institutional perspective. *Law & Policy, 19*(4), 363–414.
17. Hinz, O., & Eckert, J., (2010). The impact of search and recommendation systems on sales in electronic commerce. *Business & Information Systems Engineering, 2*(2), 67–77.
18. Hustinx, P., (2010). Privacy by design: Delivering the promises. *Identity in the Information Society, 3*(2), 253–255.
19. Jansen, B. J., (2007). Click fraud. *Computer, 40*(7), 85–86.
20. Jürgens, P., Jungherr, A., & Schoen, H., (2011). *Small Worlds with a Difference: New Gatekeepers and the Filtering of Political Information on Twitter* (Vol. 1, pp. 1–20). Association for Computing Machinery (ACM). Available at: https://www.acm.org/ (accessed on 4 August 2022).
21. Katzenbach, C., (2011). Technologies as institutions: Rethinking the role of technology in media governance constellations. In: Puppis, M., & Just, N., (eds.), *Trends in Communication Policy Research, Intellect, Bristol.* (pp. 117–138).
22. König, R., & Rasch, M., (2014). *Society of the Query Reader: Reflections on Web Search* (Vol. 1, pp. 1–23). Institute of Network Cultures, Amsterdam. Available at: https://networkcultures.org/blog/publication/society-of-the-query-reader-reflections-on-web-search/ (accessed on 4 August 2022).

23. Krishnan, S., Patel, J., Franklin, M. J., & Goldberg, K., (2014). *Social Influence Bias in Recommender Systems: A Methodology for Learning, Analyzing, and Mitigating Bias in Ratings* (Vol. 1, pp. 1–28). Available at: http://goldberg.berkeley.edu/pubs/sanjay-recsys-v10.pdf (accessed on 4 August 2022).
24. Langheinrich, M., (2001). Privacy by design – principles of privacy-aware ubiquitous systems. In: Abowd, G. D., Brumitt, B., & Shafer, S. A., (eds.), *Proceedings of the Third International Conference on Ubiquitous Computing (UbiComp 2001)*, Lecture Notes in Computer Science (LNCS), Atlanta, Georgia, Vol. 2201, pp. 273–291.
25. Lao, M., (2013). 'Neutral' search as a basis for antitrust action? *Harvard Journal of Law & Technology, 26*(2), 1–12.
26. Latzer, M., (2007). "Regulatory choice in communications governance. *Communications – The European Journal of Communication Research, 32*(3), 399–405.
27. Latzer, M., (2014). *Algorithmic Selection on the Internet: Economics and Politics of Automated Relevance in the Information Society* (Vol. 1, pp. 1–19). Research report, University of Zurich, IPMZ, Department for Media Change & Innovation.
28. Latzer, M., Hollnbuchner, K., Just, N., & Saurwein, F., (2015). The economics of algorithmic selection on the internet. In: Bauer, J., & Latzer, M., (eds.), *Handbook on the Economics of the Internet*, Edward Elgar, Cheltenham, Northampton (Vol. 1, pp. 1–25).
29. Latzer, M., Price, M. E., Saurwein, F., Verhulst, S. G., Hollnbuchner, K., & Ranca, L., (2007). *Comparative Analysis of International Co- and Self-Regulation in Communications Markets* (Vol. 1, pp. 1–20). Research report commissioned by Ofcom, ITA, Vienna.
30. Lessig, L., (1999). *Code and Other Laws of Cyberspace, Basic Books* (Vol. 1, pp. 1–25) New York, NY.
31. Lewandowski, D., (2014). "Why we need an independent index of the web. In: König, R., & Rasch, M., (eds.), *Society of the Query Reader: Reflections on Web Search* (pp. 50–58). Institute of Network Cultures, Amsterdam.
32. Lin, P., & Selinger, E., (2014). *Inside Google's Mysterious Ethics Board* (Vol. 1, pp. 1–19). Forbes. Available at: www.forbes.com/sites/privacynotice/2014/02/03/inside-googles-mysterious-ethics-board/ (accessed on 4 August 2022).

33. London Economics, (2010). *Study on the Economic Benefits of Privacy Enhancing Technologies (PETs)* (Vol. 1, pp. 1–20). Final Report to the European Commission DG Justice, Freedom and Security. available at: http://ec.europa.eu/justice/policies/privacy/docs/studies/final_report_pets_16_07_10_en.pdf (accessed on 4 August 2022).
34. Machill, M., & Beiler, M., (2007). *The Power of Search Engines: The Power of Search Engines* (Vol. 1, pp. 1–20). Herbert Von Halem Verlag, Cologne.
35. Mager, A., (2012). "Algorithmic ideology: How capitalist society shapes search engines. *Information, Communication & Society, 15*(5), 769–787.
36. Manovich, L., (2013). *Software Takes Command, Bloomsbury* (Vol. 1, pp. 1–30). New York, NY.
37. Mayer-Schönberger, V., & Cukier, K., (2013). *Big Data: Die Revolution, die Unser Leben Verändern Wird* (Vol. 1, pp. 1–30). Redline Verlag, Munich.
38. Moffat, V. R., (2009). Regulating search. *Harvard Journal of Law & Technology, 22*(2), 475–513.
39. Munson, S. A., & Resnick, P., (2010). Presenting diverse political opinions: How and how much. *Proceedings of ACM CHI 2010 Conference on Human Factors in Computing Systems 2010* (pp. 1457–1466). Atlanta, Georgia.
40. Musiani, F., (2013). Governance by algorithms. *Internet Policy Review, 2*(3). Available at: http://policyreview.info/articles/analysis/governance-algorithms (accessed on 4 August 2022).
41. Napoli, P. M., (2013). *The Algorithm as Institution: Toward a Theoretical Framework for Automated Media Production and Consumption* (Vol. 1, pp. 1–10). Paper presented at the Media in Transition Conference, MIT Cambridge.
42. Ohm, P., (2010). Broken promises of privacy: Responding to the surprising failure of anonymization. *UCLA Law Review, 57*, 1701–1777.
43. Pariser, E., (2011). *The Filter Bubble: What the Internet is Hiding from You* (Vol. 1, pp. 1–20). Penguin Books, London.
44. Pasquale, F., (2015). *The Black Box Society: The Secret Algorithms That Control Money and Information* (Vol. 1, pp. 1–20). Harvard University Press.

45. Resnick, P., Kelly, G. R., Kriplean, T., Munson, S. A., & Stroud, N. J., (2013). Bursting your (filter) bubble: Strategies for promoting diverse exposure. *Proceedings of the 2013 Conference on Computer-Supported Cooperative Work Companion* (pp. 95–100). San Antonio, Texas.
46. Rieder, B., (2005). Networked control: Search engines and the symmetry of confidence. *International Review of Information Ethics, 3,* 26–32.
47. Rietjens, B., (2006). Trust and reputation on eBay: Towards a legal framework for feedback intermediaries. *Information & Communications Technology Law, 15*(1), 55–78.
48. Saurwein, F., (2011). "Regulatory choice for alternative modes of regulation: How context matters", *Law & Policy, 33*(3), 334–366.
49. Schaar, P., (2010). Privacy by design. *Identity in the Information Society, 3*(2), 267–274.
50. Schedl, M., Hauger, D., & Schnitzer, D., (2012). A model for serendipitous music retrieval. *Proceedings of the 2nd Workshop on Context-Awareness in Retrieval and Recommendation* (pp. 10–13). Lisbon.
51. Schormann, T., (2012). *Online-Portale: Großer Teil der Hotelbewertungen ist Manipuliert* (Vol. 1, pp. 1–10). Spiegel Online. Available at: www.spiegel.de/reise/aktuell/online-portale-grosser-teil-der-hotelbewertungenist-manipuliert-a-820383.html (accessed on 4 August 2022).
52. Schulz, W., Held, T., & Laudien, A., (2005). "Search engines as gatekeepers of public communication: Analysis of the German framework applicable to internet search engines including media law and anti-trust law. *German Law Journal, 6*(10), 1418–1433.
53. Senecal, S., & Nantel, J., (2004). The influence of online product recommendation on consumers' online choice. *Journal of Retailing, 80*(2), 159–169.
54. Sinclair, D., (1997). Self-regulation versus command and control? Beyond false dichotomies. *Law & Policy, 19*(4), 529–559.
55. Somaiya, R., (2014). *How Facebook is Changing the Way Its Users Consume Journalism* (Vol. 1, pp. 1–10). New York Times. Available at: www.nytimes.com/2014/10/27/business/media/how-facebook-ischanging-the-way-its-users-consume-journalism.html (accessed on 4 August 2022).

56. Steiner, C., (2012). *Automate This: How Algorithms Came to Rule Our World* (Vol. 1, pp. 1–10). Penguin Books, New York, NY.
57. Van, D. A., (2012). "The algorithms behind the headlines. *Journalism Practice, 6*(5, 6), 648–658.
58. Wallace, J., & Dörr, K., (2015). "Beyond traditional gatekeeping. How algorithms and users restructure the online gatekeeping process. *Conference Paper, Digital Disruption to Journalism and Mass Communication Theory*, Brussels.
59. Wittel, G. L., & Wu, S. F., (2004). On attacking statistical spam filters. *Proceedings of the First Conference on Email and Anti-Spam (CEAS)*. Available at: http://pdf.aminer.org/000/085/123/on_attacking_statistical_spam_filters.pdf (accessed on 4 August 2022).
60. Zittrain, J., & Palfrey, J., (2008). "Internet filtering: The politics and mechanisms of control. In: Deibert, R., Palfrey, J., Rohozinski, R., & Zittrain, J., (eds.), *Access Denied: The Practice and Policy of Global Internet Filtering* (pp. 29–56). MIT Press, Cambridge.

INDEX

Printed in the United States
by Baker & Taylor Publisher Services